U.S.NRC
United States Nuclear Regulatory Commission

Protecting People and the Environment

NUREG/CP-0195

I0482772

Proceedings of the Workshop on Engineered Barrier Performance Related to Low-Level Radioactive Waste, Decommissioning, and Uranium Mill Tailings Facilities

Held at the U.S. Nuclear Regulatory Commission Headquarters, Rockville, MD
August 3-5, 2010

Office of Nuclear Regulatory Research

Office of Federal and State Materials and Environmental Management Programs

AVAILABILITY OF REFERENCE MATERIALS
IN NRC PUBLICATIONS

NRC Reference Material

As of November 1999, you may electronically access NUREG-series publications and other NRC records at NRC's Public Electronic Reading Room at http://www.nrc.gov/reading-rm.html. Publicly released records include, to name a few, NUREG-series publications; *Federal Register* notices; applicant, licensee, and vendor documents and correspondence; NRC correspondence and internal memoranda; bulletins and information notices; inspection and investigative reports; licensee event reports; and Commission papers and their attachments.

NRC publications in the NUREG series, NRC regulations, and *Title 10, Energy*, in the Code of *Federal Regulations* may also be purchased from one of these two sources.
1. The Superintendent of Documents
   U.S. Government Printing Office
   Mail Stop SSOP
   Washington, DC 20402-0001
   Internet: bookstore.gpo.gov
   Telephone: 202-512-1800
   Fax: 202-512-2250
2. The National Technical Information Service
   Springfield, VA 22161-0002
   www.ntis.gov
   1-800-553-6847 or, locally, 703-605-6000

A single copy of each NRC draft report for comment is available free, to the extent of supply, upon written request as follows:
Address: U.S. Nuclear Regulatory Commission
           Office of Administration
           Publications Branch
           Washington, DC 20555-0001
E-mail: DISTRIBUTION.SERVICES@NRC.GOV
Facsimile: 301-415-2289

Some publications in the NUREG series that are posted at NRC's Web site address http://www.nrc.gov/reading-rm/doc-collections/nuregs are updated periodically and may differ from the last printed version. Although references to material found on a Web site bear the date the material was accessed, the material available on the date cited may subsequently be removed from the site.

Non-NRC Reference Material

Documents available from public and special technical libraries include all open literature items, such as books, journal articles, and transactions, *Federal Register* notices, Federal and State legislation, and congressional reports. Such documents as theses, dissertations, foreign reports and translations, and non-NRC conference proceedings may be purchased from their sponsoring organization.

Copies of industry codes and standards used in a substantive manner in the NRC regulatory process are maintained at—
   The NRC Technical Library
   Two White Flint North
   11545 Rockville Pike
   Rockville, MD 20852-2738

These standards are available in the library for reference use by the public. Codes and standards are usually copyrighted and may be purchased from the originating organization or, if they are American National Standards, from—
   American National Standards Institute
   11 West 42nd Street
   New York, NY 10036-8002
   www.ansi.org

Legally binding regulatory requirements are stated only in laws; NRC regulations; licenses, including technical specifications; or orders, not in NUREG-series publications. The views expressed in contractor-prepared publications in this series are not necessarily those of the NRC.

The NUREG series comprises (1) technical and administrative reports and books prepared by the staff (NUREG-XXXX) or agency contractors (NUREG/CR-XXXX), (2) proceedings of conferences (NUREG/CP-XXXX), (3) reports resulting from international agreements (NUREG/IA-XXXX), (4) brochures (NUREG/BR-XXXX), and (5) compilations of legal decisions and orders of the Commission and Atomic and Safety Licensing Boards and of Directors' decisions under Section 2.206 of NRC's regulations (NUREG-0750).

DISCLAIMER: Where the papers in these proceedings have been authored by contractors of the U. S. Government, neither the U.S. Government nor any agency thereof, nor any U.S. employee makes any warranty, expressed or implied, or assumes any legal liability or responsibility for any third party's use or the results of such use, of any information, apparatus, product, or process disclosed in these proceedings, or represents that its use by such third party would not infringe privately owned rights. The views expressed in these proceedings are not necessarily those of the U. S. Regulatory Commission.

NUREG/CF-0195

United States Nuclear Regulatory Commission

Protecting People and the Environment

Proceedings of the Workshop on Engineered Barrier Performance Related to Low-Level Radioactive Waste, Decommissioning, and Uranium Mill Tailings Facilities

Held at the U.S. Nuclear Regulatory Commission Headquarters, Rockville, MD
August 3-5, 2010

Manuscript Completed: June 2011
Date Published: August 2011

Prepared by:
T. J. Nicholson and H.D. Arlt

Office of Nuclear Regulatory Research

Office of Federal and State Materials and Environmental
   Management Programs

DISCLAIMER

This report is not a substitute for U.S. Government regulations, and compliance with the information and guidance provided is not required. The technical approaches, software, and methods described in these conference proceedings are provided for information only. Publication of these proceedings does not necessarily constitute Federal agency approval or agreement with the information contained herein. Use of product or trade names is for identification purposes only and does not constitute endorsement or recommendation for use by any Federal agency.

The views expressed in these proceedings are those of the individual authors and do not necessarily reflect the views or policies of the U.S. Nuclear Regulatory Commission (NRC) and the other participating Federal agencies.

NUREG/CP-0195 has been
reproduced from the best available copy.

ABSTRACT

NRC's Offices of Nuclear Regulatory Research (RES) and Federal and State Materials and Environmental Management Programs (FSME) organized this *Workshop on Engineered Barrier Performance Related to Low-Level Radioactive Waste, Decommissioning, and Uranium Mill Tailings Facilities*. The workshop was held August 3–5, 2010 at the U.S. Nuclear Regulatory Commission (NRC) Headquarters Auditorium, 11545 Rockville Pike, Rockville, Maryland. The Workshop was coordinated with the States (i.e., Texas, South Carolina, Utah, Colorado, Washington, and New York), Tribal Nations (Navajo, Umatilla and Nez Perce), and Federal agencies (e.g., U.S. Department of Energy [DOE], U.S. Environmental Protection Agency [EPA], U.S. Department of Agriculture's Agricultural Research Service [USDA/ARS], U.S. Geological Survey [USGS], and DOE National Laboratories). The workshop technical topics focused on engineered surface covers and bottom liners designed to isolate waste by impeding surface-water infiltration into the waste systems and mitigating the migration of contaminants from the waste disposal site. Topics included engineered barrier performance, modeling, monitoring, and regulatory experiences at low-level radioactive waste, decommissioning, and uranium mill tailings sites. The workshop objectives included: (1) facilitation of communication among Federal and State staff and contractors and selected experts on current engineered barrier issues and technical and regulatory experiences; (2) discussion of lessons learned and approaches for monitoring and modeling; (3) preparation of recommendations to address maintenance of engineered barrier performance over time; and (4) identification of topics for future research and the potential need to update technical guidance. Recommendations and insights given during session presentations, panel debates, and the discussions that followed were documented by the session reporters and are included in this report.

TABLE OF CONTENTS

ABSTRACT iii

CONTENTS v

ACKNOWLEDGEMENTS xi

EXECUTIVE SUMMARY xiii

1. INTRODUCTION 1-1

1.1 Background to the Engineered Barrier Performance Workshop 1-2
1.2 Engineered Barrier Performance Workshop 1-2
1.3 Overview of Engineered Barrier Types, Components, Performance, and Regulatory
    Compliance Criteria 1-3
1.3.1 Engineered Barrier Systems, Components and Function 1-3
1.3.2 Regulatory Compliance Criteria 1-5
1.3.3 Engineered Barrier System Performance 1-6

2. SESSION 1A: EXPERIENCE OF THE STATES IN REGULATING FACILITIES
         INVOLVING ENGINEERED COVERS AND LINERS
         Session Chairs: Stephen Salomon, NRC/FSME and Susan Jablonski, TCEQ,
         State of Texas, Technical Reporter: Douglas Mandeville, NRC/FSME 2-1

2.1 Motivation 2-2
2.2 Background 2-2
2.3 Summary of Panel Discussion 2-2
2.4 Summary and Recommendations 2-3
2.5 Extended Abstracts 2-4
2.5.1 *Modeling and Monitoring of Barrier Performance for the Planned Texas Low-Level
        Radioactive Waste Disposal Facility*
        by Susan Jablonski, Peter Lodde, and Abel Porras 2-5
2.5.2 *Utah Clive Low-Level Radioactive Waste Facility*
        by Loren Morton 2-13
2.5.3 *Overview of the Performance and Use of Engineered Barriers at the Barnwell
        LLRW Disposal Site*
        by Susan E. Jenkins 2-21
2.5.4 *Washington State's Experience with Decommissioning and Evaluation of Cover
        Designs for Low-Level Radioactive Waste and Uranium Mill Tailings Facilities*
        by Gary Robertson 2-24
2.5.5 *State of Colorado Experience with Waste Repository Covers and Caps*
        by Lawrence J. Bruskin and Steve Tarlton 2-31
2.5.6 *The Navajo Nation Environmental Protection Agency Experience with
        Former Uranium Mill Sites*
        by Stephen A. Austin 2-36
2.5.7 *New York State Department of Environmental Conservation's Experience with
        Solid and Hazardous Waste and Low-Level Radioactive Waste Containment
        Systems – A Lessons Learned Perspective*
        by Robert Phaneuf, Tim Rice, Dave O'Hehir, John Mitchell and Jaime Lang 2-38

3. SESSION 1B: FEDERAL AGENCIES AND DOE NATIONAL LABORATORIES
    Session Chairs: Jacob Philip, NRC/RES and Brian Andraski, U.S. Geological
    Survey, Technical Reporter: George Alexander, NRC/FSME 3-1

3.1 Motivation 3-2
3.2 Background 3-2
3.3 Challenges/Summaries of Session Presentations 3-2
3.4 Recommendations 3-4
3.5 Extended Abstracts 3-4
3.5.1 *Waste Isolation and Contaminant Migration – Tools and Techniques for Monitoring
    the Saturated Zone–Unsaturated Zone-Plant-Atmosphere Continuum*
    by Brian J. Andraski and David A. Stonestrom 3-5
3.5.2 *USACE Experience with HTW Containment Systems*
    by Kevin Pavlik 3-9
3.5.3 *Design Basis of UMTRA Disposal Cells*
    by Rich Bush, John Elmer and Greg Smith 3-10
3.5.4 *EPA's Review of its Regulatory Requirements for Uranium and Thorium Mill
    Tailings: 40 CFR Part 192*
    by Loren Setlow 3-14
3.5.5 *Investigations Supporting Performance Verification of Engineered Barrier Systems*
    by Joel Hubbell 3-15
3.5.6 *Savannah River Site E-Area Low-Level Waste Facility Subsidence Studies
    SRNL-MS-2010-00128, Revision 1*
    by Mark Phifer 3-16
3.5.7 *DOE Overview*
    by Martin Letourneau 3-26

4. SESSION 2: DEGRADATION PROCESSES AND PERFORMANCE EVOLUTION OF
    ENGINEERED BARRIERS
    Session Chairs: Craig Benson, University of Wisconsin/CRESP and W. Jody
    Waugh, S.M. Stoller LLC;Technical Reporter: Brooke Traynham, NRC/FSME ... 4-1

4.1 Motivation 4-2
4.2 Background 4-2
4.3 Challenges in Prediction of Cover Performance 4-2
4.4 Recommendations 4-3
4.5 Extended Abstracts 4-4
4.5.1 *DOE Experience with Cover Degradation Processes, Design Improvements, and
    Cover Renovation for Uranium Mill Tailings Disposal Cells*
    by W. J. Waugh 4-5
4.5.2 *Processes that Alter the Structure of Soils and Their Effects on the Performance
    of Covers*
    by Craig H. Benson 4-12
4.5.3 *Assessment of the Erosional Stability of Encapsulation Caps and Covers at the
    Millennial Timescale: Current Capabilities, Research Issues and Operational
    Needs*
    by Gary R. Willgoose 4-16
4.5.4 *Ecological Processes and Changes in the Performance of Covers*
    by Steven O. Link 4-20

4.5.5 *Degradation Processes and Changes in the Performance of Geomembranes*
    by R. Kerry Rowe 4-23

5. SESSION 3: EXPERIENCE WITH MONITORING DEVICES AND SYSTEMS
    USED TO MEASURE PERFORMANCE
    Session Chairs: William Albright, Desert Research Institute/UNV and
    Craig Benson, University of Wisconsin/CRESP,
    Technical Reporter: Robert Johnson, NRC/FSME 5-1

5.1 Motivation 5-2
5.2 Background 5-2
5.3 Contemporary Issues 5-2
5.4 Recommendations 5-3
5.5 Extended Abstracts 5-4
5.5.1 *In Search of the Perfect Cap: 15 Years of Performance Data from the Prototype
    Hanford Barrier*
    by A. L. Ward 5-5
5.5.2 ACAP: *Monitoring Cover Performance and Changes in Performance with
    Drainage Lysimeters, Instruments, and Exhumations*
    by William H. Albright and Craig H. Benson 5-7
5.5.3 *Monitoring Contaminant Strategies: Tools, Techniques, Methodologies and Modeling
    Approaches*
    by Timothy J. Gish, Audrey K. Guber, and Yakov A. Pachepsky 5-9
5.5.4 *Aerial Remote Sensing as a Component of Closure Cap Monitoring*
    by John B. Gladden 5-12
5.5.5 *Differential Settlement and its Importance on the Performance of Cover Systems
    at Radiological Waste Disposal Facilities*
    by Robert C. Bachus and John F. Beech 5-15

6. SESSION 4: MODELING EXPERIENCES IN PERFORMANCE ASSESSMENT
    AND EVALUATION OF PERFORMANCE MONITORING
    Session Chairs: David Esh, NRC/FSME and Thomas Nicholson, NRC/RES,
    Technical Reporter: Christopher Grossman, NRC/FSME 6-1

6.1 Motivation 6-2
6.2 Background 6-2
6.3 Challenges and Experiences in Performance Modeling 6-2
6.4 Recommendations 6-4
6.5 Extended Abstracts 6-4
6.5.1 *Development of an Integrated Probabilistic Model of Radiological Fate and
    Transport in an Engineered Cover*
    by John Tauxe 6-5
6.5.2 *Practical Considerations for Modeling and Monitoring of Engineered Barrier
    Performance*
    by Roger Seitz 6-10
6.5.3 *Near-Term Hydrological Performance Modeling of Final Covers*
    by Craig H. Benson 6-14

6.5.4 *Simulating the Long-Term Performance of Engineered Barriers in Arid Environments with the STOMP Sparse Vegetation Evapotranspiration Model*
by A. L. Ward 6-17

6.5.5 *Effects of Plant Succession on the Functioning of Engineered Covers and Modeling of Long-Term Successional Impacts Using the EDYS Ecological Simulation Model*
by Terry McLendon 6-20

6.5.6 *Applications of Thermal Remote Sensing for Multi-Scale Monitoring of Evapotranspiration*
by William P. Kustas and Martha C. Anderson 6-25

7. SESSION 5: EXPERIENCE WITH MODEL SUPPORT AND MULTIPLE LINES OF EVIDENCE TO GAIN CONFIDENCE IN LONG-TERM PERFORMANCE
Session Chairs: Hans Arlt, NRC/FSME and George Alexander, NRC/FSME
Technical Reporter: Brooke Traynham, NRC/FSME 7-1

7.1 Background and Motivation 7-2
7.2 Highlights of Session Presentations 7-2
7.3 Recommendations/Insights from Panel Discussion 7-4
7.4 Extended Abstracts 7-6
7.4.1 *Overview of Model Support (for Engineered Barriers)*
by Dave W. Esh 7-7
7.4.2 *Activities that Support the Scientific Credibility of Radioactive Waste System Performance Models*
by Abraham Van Luik 7-12
7.4.3 *Geomembrane Performance in Landfill Cover Systems*
by George R. Koerner 7-14
7.4.4 *A Role for Natural Analogs in the Design and Long-Term Performance Evaluation of Earthen Covers for Uranium Mill Tailings*
by William J. Waugh 7-18
7.4.5 *Arid Soil Evolution and Pedologic Development: Process Considerations and Applications to Engineered Barrier Design*
by Todd G. Caldwell, Michael H. Young, and Eric V. McDonald 7-24
7.4.6 *Historical Studies of Moisture Contents and Infiltration Through Compacted Clay Radon Barriers at Rock Covered UMTRA Disposal Cells*
by Kent Bostick 7-28

8. SESSION 6: RECOMMENDATIONS ON ASSESSING ENGINEERED BARRIER PERFORMANCE, IDENTIFYING FUTURE RESEARCH NEEDS, AND DISCUSSING EXISTING GUIDANCE
Session Chairs: Thomas Nicholson, NRC/RES and Hans Arlt, NRC/FSME
Technical Reporters: Mark Fuhrmann, NRC/RES and Allen Gross, NRC/FSME
.... 8-1

8.1 Overview and Insights 8-2
8.2 Observations 8-2
8.2.1 Principal Observations 8-3
8.2.2 Additional General Observations 8-4
8.2.3 Specific Observations 8-4
8.2.4 Monitoring and Modeling Observations 8-5

8.2.5   Model Support Observations 8-6

8.2.6   Processes and Performance Observations 8-7

8.3 NRC Staff Observations and Recommendations 8-3

APPENDIX A: Workshop Agenda A-1

APPENDIX B: Workshop Participants B-1

APPENDIX C: Acronyms C-1

APPENDIX D: Bibliography D-1

APPENDIX E: Electronic Information Sources E-1

ACKNOWLEDGEMENTS

The concept, planning and execution of this workshop, and the development and documentation of these proceedings were achieved by an organizing committee composed of Federal and non-Federal volunteers. The organizing committee consisted of: William Albright, Desert Research Institute/University of Nevada; George Alexander, NRC/FSME; Brian Andraski, U.S. Geological Survey; Hans Arlt, NRC/FSME; Craig Benson, University of Wisconsin/CRESP; Mark Fuhrmann, NRC/RES; Susan Jablonski, State of Texas; Thomas Nicholson, NRC/RES; Jacob Philip, NRC/RES; Stephen Salomon, NRC/FSME; Loren Setlow U.S. Environmental Protection Agency; and W. Jody Waugh, S.M. Stoller LLC.

The organizing committee is grateful for the support of David Esh, NRC/FSME; Allen Gross, NRC/FSME; Christopher Grossman, NRC/FSME; Robert Johnson, NRC/FSME; Douglas Mandeville, NRC/FSME; and Brooke Traynham, NRC/FSME; who also served as Session Chairs and Reporters.

We are appreciative of the NRC Program Support Staff: Danita Stenberg, Sarah Achten, Tarsha Moon, John Wucher and Nathan White, Khai Tran, Matt Williams; and NRC Management: Chris McKenney, NRC/FSME and Gary Simpler (Security) who greatly facilitated the logistics for this large workshop.

Finally, we are indebted to all of the 35 presenters and 15 panelists, who are listed in the table of contents, for their efforts to prepare their presentations, to travel to Rockville, MD, and to present and discuss them. We are also indebted to the approximately 140 workshop attendees who actively engaged in questioning the speakers and panelists during the panel discussions. Together these presenters and panelists generated significant information and observations (as noted in the extended abstracts and session summaries documented in these proceedings) which made for a very successful workshop and report.

EXECUTIVE SUMMARY

The motivation for convening this engineered barrier performance workshop was to identify and present new information from field studies, monitoring programs, modeling studies and regulatory experiences related to engineered covers and liners related to low-level radioactive waste, decommissioning, and uranium mill tailings. The workshop objectives were: (1) to facilitate communication and exchange of technical and regulatory experiences among selected experts on current engineered barrier issues, including academics, Federal and State staff, and contractors; (2) discuss lessons learned and approaches for monitoring and modeling; preparation of recommendations to address maintenance of engineered barrier performance over time; and (3) identify topics for future research and the potential need to update technical guidance. In particular, the technical issues focused on degradation processes and changing performance of engineered barriers, monitoring (short-term), model support (long-term), and modeling of processes within the barriers, especially engineered surface covers in addition to practical examples of performance failures and successes based on field observations.

An organizing committee comprised of Federal and State regulators, academic and industry experts was formulated. The organizing committee identified the specific technical issues and appropriate experts as invited speakers and panelists. These participants included Federal and State staff and contractors, selected academic and industry experts, representatives from Tribes, and NRC technical staff and management (please see Workshop Participants list). The public was welcomed to attend and observe the proceedings of this Category 1 Public Meeting which was posted on the NRC Public Meeting Website. About 140 workshop registrants were in attendance. In addition, several hundred viewers observed the workshop proceedings remotely via "WebStreaming" which was also identified on the NRC Public Meeting Website.

This workshop was coordinated with the States, Tribal Nations, and Federal Agencies. The organizing committee (please see Acknowledgements for a listing of members) developed a tentative agenda that included technical session themes with a series of questions to help focus the presentations and panel discussions. Speakers and panelists were requested to address these themes and questions. The organizing committee developed a list of candidate presenters based upon these technical themes and questions identified in the tentative agenda.

In the introductory session, the workshop objectives, technical themes, and topics were presented. This was followed by a presentation on descriptions of the various engineered barrier types by function and design and, finally, an overview of NRC's experience with engineered barrier performance in research, licensing, and regulatory compliance.

Technical Sessions included:
- *Session 1A* – Experience of the States in Regulating Facilities Involving Engineered Covers and Liners. States and a Native Tribe presented an overview of their research activities and findings with an emphasis on practical insights on monitoring, modeling, and confirming short- and long-term performance of engineered systems. During the panel discussion, common themes emerged including a) limitations to current technologies and the uncertainties of changing regulations; b) site specific characteristics can create difficulty in maintaining a consistent and transparent approval process; c) site specific characteristics will also affect the approach to engineered cover design; d) full cost accounting of long-term maintenance and monitoring is needed; and e) meetings such as this workshop are helpful in exchanging ideas from specialists in a wide variety of disciplines and therefore should be considered as an annual event.

- ***Session 1B*** – Federal Agencies and DOE National Laboratories presented an overview of their research activities. Significant observations and recommendations from the panel discussions included a) the need for monitoring adjacent to the waste containment facility; b) the importance of vadose zone monitoring; c) the importance of scale effects for input parameters in PA; and d) the continued use of results from PA monitoring and modeling to identify and prioritize research needs.

- ***Session 2*** – Degradation Processes and Performance Evolution of Engineered Barriers and Covers. Major discussion topics included physical, chemical, and biological degradation processes. These include freeze-thaw, wet-dry cycle, differential settlement, retention of borrow soil structure (peds or clods) during construction, UV degradation, thermal degradation, erosion, fire, and pedogenesis. Chemical processes include oxidation of geosynthetic materials and cation exchange mechanisms in sodium bentonites. Biological processes include unanticipated ecological consequences of designs that, by creating habitat for deep-rooted plants, burrowing animals, and soil microorganisms, can alter soil hydraulic properties.

  The challenge of predicting cover performance in the context of the aforementioned degradation processes is confounded by the need to infer long-term conditions from short-term monitoring data, difficulty in scaling from bench to field, data gaps, developing/screening future scenarios, standardize future intruder scenarios, unforeseen ecological consequences, and the tendency of root intrusion and soil development to alter hydraulic properties. Understanding the range of time scales over which different degradation processes operate is important when characterizing degradation processes. Timescales of 100s to 1000s of years need to be considered as well as the spatial heterogeneity of the system.

- ***Session 3*** – Experience with Monitoring Devices and Systems Used to Measure Performance. Scientists and engineers face the challenge of understanding and predicting the performance of entire containment systems, not just individual components. This requires monitoring systems designed to evaluate component interactions and changes to the system that result from environmental influences. Performance monitoring should include direct measurement of the critical parameter (e.g. percolation through a cover) as well as data from supplemental instruments to understand processes and changes to the containment structure. Too often, the most important performance parameter is calculated from peripheral data or estimated from modeling. Not only is direct measurement critical by itself (when possible), but it provides accurate context for interpretation of the status and changes to the containment system.

  Monitoring should be driven by the purpose of the project. Monitoring systems should be designed to understand processes and identify precursors of problems. Remote sensing can provide an alternate view of a containment system. Advantages include complete coverage, detection of factors not seen by the eye, automation, and potentially lower cost. Remote sensing can focus evaluations by flagging issues that might have been missed by visual surveillance.

- ***Session 4*** – Modeling Experiences in Performance Assessment and Evaluation of Performance Monitoring. Workshop participants provided valuable insights and experiences with respect to the modeling of surface barriers and waste disposal liners. For waste disposal systems, integration of processes in modeling is essential because the processes are integrated in the real world. Influence diagrams and conceptual drawings can be useful in refining the scope of modeling. Graded and iterative approaches, common to performance assessments, have been found to be useful in the modeling of engineered covers and liners. Processes determining cover performance can be non-linear and typically uncertainty is large, especially for longer-term assessments. Therefore, integration is extremely important and cover modeling needs to take into account the uncertainty in performance predictions. The models may have different forcing functions, and they need the capability to incorporate the influence of those different forcing functions. Model completeness is difficult to assess. At a minimum, models must include water flow and transport, soil loss by erosion, surface water runoff, and bioturbation by animals and plants.

Ecological modeling has been progressing, and is a very important component to assessing the performance of engineered covers. Roots penetrate natural barriers, and plant succession occurs everywhere. Two common aspects of the succession are an increase in vegetation structure and an increase in the relative amounts of woody plants. Both of these aspects have profound implications to the function and performance of engineered barriers.

Long timeframes are very difficult to account for in modeling of engineered covers and liners. Participants identified processes, especially hydrologic, that require high frequency data (e.g. daily or hourly) in order to capture observed responses. In addition, simulations of long-term performance must incorporate a variety of potential degradation mechanisms such as extreme weather events, freeze thaw cycling, fire, and bio-intrusion. Modeling coupled processes with substantially different response times can be computationally and intellectually challenging, but progress is being made.

- **Session 5** – Experience with Model Support and Multiple Lines of Evidence to Gain Confidence in Long-Term Performance. Model support is an important element in the application of engineered barriers to limit risk and helps ensure the likelihood of making a good decision based on performance projections. Best practices include: Multiple lines of evidence, direct observations, risk informed, expert elicitation, accelerated experiments, natural analogs, and support for full range of expected conditions. Natural analogs may provide some data for very long term, however they can have a confirmation bias (other analogs may have weathered away) and the exposure conditions are unknown and can be highly dynamic. Complexity of the model should not be greater than the model support. Evaluation of features, events, and processes (FEPs) provides a way to identify relevant and non-redundant elements and scenarios to be analyzed. Those that pass through screening test (credible and likely to occur) will be included in model. Performance margin analysis includes less conservative alternative conceptual models. Both natural and anthropogenic analogs can support model development. Independent technical review process provides peer review recommendations. Performance confirmation monitoring and testing is conducted to evaluate the adequacy of the information used to demonstrate safety by confirming that system is operating as intended. Competing objectives and goals require tradeoffs which should be explicitly acknowledged.

The role of natural analogs in the design and long-term performance evaluation of earthen covers for uranium mill tailings, and the ways they may be of value as a line of evidence in model support were discussed in depth in this session. In the long term, the ecology of a cover will change in ways that cannot be accurately predicted by models alone. Ecosystem engineering paradigm is a useful way to view the cover system since earthen covers are engineered ecosystems that will be greatly influenced by surrounding ecosystem. Analogs provide tangible evidence and clues for understanding these changes, designing field experiments, and developing and screening scenarios for performance modeling. Some examples: Paleoclimate data provides more local data than regional climate change models (finer scale data). Fire chronosequences provides evidence of recover after a fire disturbance. Side slope stability can be examined with an ancient hill analog or glacial debris flow (long term erosion protection). Pedogenic carbonates provide analog of a capillary barrier on the order of 10,000 yrs. Further session topics discussed included soil evolution and pedologic development applicable to engineered barrier designs. Soil development will be a function of climate, parent material, topography, biology, and time. Pedogenic features will affect hydraulic processes and plant communities at various scales, both spatially and temporally. The modern landscape provides a record of soil evolution pathways that can support long term assessments of covers. Hydraulic properties of soil are a function of both biotic and abiotic processes – roots growth, faunal burrowing, nutrient cycling, and resource translocation. Soils undergo a development of macropores with time, transitioning from micropore dominated soils. There is a predictable decrease in the permeability of the upper horizon when starting with lose wind-blown sediments, resulting in more water being retained at the surface. Young soils have no development implying sandy texture with high infiltration while old deposits are clay rich. Incipient soil formation - short term soil development processes relevant to cover evolution. Pedogenesis for a compacted clay, or engineered silt loam will likely evolve much faster. Vegetation distribution and size is largely a function of soil grain size and composition. Feedback between soil and vegetation will influence: canopy height/volume, shrub abundance, rooting depth and lateral spread, plant type.

- **Session 6** – Recommendations on Assessing Engineered Barrier Performance, Identifying Future Research Needs, and Improving Guidance Documents. Over the last few years, research results from various organizations, including NRC/RES, have raised technical questions regarding performance of engineered surface barriers, and assumptions in performance assessments of these facilities. All participants were very interested in obtaining a greater understanding of the processes that effect short-term and long-term performance of engineered barriers. Two of the most consistent themes to come from the workshop participants were 1) more communication and exchange was desired and needed between all those involved with engineered barriers, and 2) much more effort must be made to monitor and collect data on engineered barrier performance so as to know the record of performance and better understand the key processes effecting performance. These insights were voiced by all engaged participants and repeated throughout the three-day workshop.

At the end of each working session, a technical panel of the presenters and selected experts responded to questions that had been formulated earlier in order to help focus discussion on pertinent technical issues. Technical reporters captured significant insights and recommendations; a few of the more major insights and recommendations are as follow:

- Engineered barriers have only been used for a few decades to isolate hazardous and radioactive waste from human populations. It is therefore not surprising that research and documentation continues on the overall assessment of the performance of engineered barrier systems for various lengths of time and for very diverse sites. Therefore, individuals and groups involved with engineered barrier performance need to communicate more frequently and coordinate efforts.

- As-built properties of for each component of an engineered barrier cannot be expected to remain indefinitely constant if exposed to the forces and processes of the environment. Designed-based criteria for such features are inappropriate. Actual performance needs to monitored or validated by some supporting information.

- Until the interrelationship between these processes and barrier performance is better understood, more data will need to be collected and analyzed. The level of monitoring and data collection should be risk informed and commensurate with the amount of performance assumed. Data collected should be relevant to the time period of reliance (performance) on engineered barrier. If performance is assumed to be relatively long-term, e.g., engineered surface covers designed for waste-incidental-to-reprocessing (WIR), data collection should help confirm main assumptions and be more in the form of robust and site-specific model support. Based on the information presented during the workshop, NRC staff would expect increased data from the following significant areas:

    • Increased interest in the long-term application of evapotranspiration and geomembranes to minimize water infiltration through deposited waste.

    • Increased interest in pedogenic processes, biotic activities, and bioturbation occurring within the cover soils and drainage layers.

- A total systems approach to monitoring and modeling is recommended. Engineered covers and liners should not be looked at in isolation from the entire waste disposal system, but instead be incorporated into a graded and iterative approach to identify significant processes and components requiring further detailed modeling and/or monitoring. This approach is in alignment with NRC's general defense-in-depth approach for performance.

INTRODUCTION

1.1 Background to the Engineered Barrier Performance Workshop

The concept of engineered barriers to isolate waste began to take hold roughly fifty years ago in the United States, and only recently that sufficient data has been gathered to begin to assess performance of engineered surface covers and bottom liners. Research results from various organizations, such as the States, the National Academies, DOE, EPA, NRC, and many others, have allowed a better understanding of engineered barrier performance. These include: the National Research Council of the National Academies publishing "Assessment of the Performance of Engineered Waste Containment Barriers" in 2007; DOE sponsoring the "Alternative Landfill Cover Demonstration" project on conventional and alternative cover designs; EPA conducting the "Alternative Cover Assessment Program" (ACAP) to evaluate the performance of alternative landfill covers beginning in 1998; and DOE hosting a "landfill" workshop in 2008 involving various DOE organizations, National Laboratories, NRC, and Academia to assess the state-of-the-practice of engineered covers. Previous workshops and meetings on engineered surface covers and bottom liners have produced informed recommendations and useful insights. However, there was a need for a broader group of cooperating organizations to focus on longer time periods, covering a wider range of environmental conditions, waste forms, and field experiences.

1.2 Engineered Barrier Performance Workshop

The workshop was held August 3–5, 2010 at the U.S. Nuclear Regulatory Commission (NRC) Headquarters Auditorium, 11545 Rockville Pike, Rockville, Maryland. The Workshop was coordinated with the States (i.e., Texas, South Carolina, Utah, Colorado, Washington, and New York), Native American Indian Tribes, and Federal agencies (e.g., U.S. Department of Energy [DOE], U.S. Environmental Protection Agency [EPA], DOE/ARS, U.S. Geological Survey [USGS], and DOE National Laboratories). The workshop technical topics focused on engineered surface covers and bottom liners designed to isolate waste by impeding surface-water infiltration into the waste systems and mitigating the migration of contaminants from the waste disposal site. Topics were divided into technical sessions and included engineered barrier performance, modeling, monitoring, and regulatory experiences at low-level radioactive waste, decommissioning, and uranium mill tailings sites. The workshop objectives included: (1) Facilitation of communication among Federal and State staff and contractors and selected experts on current engineered barrier issues and technical and regulatory experiences. (2) Discussion of lessons learned and approaches for monitoring and modeling. (3) Preparation of recommendations to address maintenance of engineered barrier performance over time. (4) Identification of topics for future research and the potential need to update technical guidance.

In the introductory session, the workshop objectives, technical themes, and topics were presented. This was followed by a presentation on descriptions of the various engineered barrier types by function and design and, finally, an overview of NRC's experience with engineered barrier performance in research, licensing, and regulatory compliance. At the end of each working session, a technical panel of the presenters and selected experts responded to questions. Technical reporters captured significant insights and recommendations from these panelists that were reported during the final session. At the end of each day's sessions, the public was provided an opportunity to make comments or to provide questions. Recommendations and insights given during session presentations, panel debates, and the discussions that followed were documented by technical reporters and included in this report.

This workshop was fortunate to have many organizations and many individuals with extensive expertise participate. It was a unique opportunity for organizations and many individuals to present their knowledge and learn what others have discovered in this area. One of the main objectives of this workshop, to facilitate the exchange of information and ideas as it relates to engineered barrier performance, was positively accomplished. Participants included invited speakers and panelists, Federal and State staff and contractors, staff from Tribal Nations, selected experts, representatives from Tribes, and NRC technical staff and management. The public was welcomed to attend and observe the proceedings of this Category 1 Public Meeting. About 140 workshop registrants were in attendance. In addition, several hundred viewers observed the workshop proceedings remotely via "WebStreaming."

1.3 Overview of Engineered Barrier Types, Components, Performance, and Regulatory Compliance Criteria

1.3.1 Engineered Barrier Systems, Components and Functions

The components of engineered barrier systems (EBS) may include liners, covers or caps, and/or lateral barriers or walls, and may use a variety of natural material such as aggregates, soil, or clay, and synthetic, cementitious, and bituminous materials including polyethylenes, fabrics, mortar, and asphalt. The regulatory disposal requirements, and types of EBS chosen, depend on the waste type. Of all the components from the EBS, the engineered surface barrier, or cover, is the most commonly used barrier and often considered to be one of the most important components. Since engineered surface covers can be significant barriers, they may provide reasonable assurance that one or more performance objectives will be met. For example, covers may greatly contribute to performance so that the following objectives for safe near-surface disposal of LLW as set in 10 CFR Part 61 can be reached: provide physical stabilization of the site (10 CFR 61.44), minimize infiltration and slow degradation of the stabilized wasteform (10 CFR 61.41), and provide an intruder deterrent (10 CFR 61.42).

There are three main functions for covers. The physical presence of an engineered surface cover can retard various environmental processes and enhance the waste's isolation. Thicker covers can provide protection from external radiation while concurrently hindering all but the deepest roots from extracting contaminants from the subsurface and releasing them at the more environmentally active surface. Burrowing animals may damage the cover but are usually impeded from obtaining direct contact by the depth at which the waste is located. For example, the engineered surface barrier for the grout-filled tanks currently being considered at the Hanford Site in the State of Washington has a thickness of 5 m [15 ft] while the top of the grout-filled tanks are currently covered with approximately 2 m [7 ft] of backfill. Engineered surface covers are expected to deter future inadvertent intrusion by man including possible excavation during construction of structures. Surface covers can provide physical stabilization especially at uranium mill tailings sites. NUREG-1623 provides guidelines on designing erosion resistant covers for long-term stabilization.

Another function of engineered surface barriers is demonstrated at radioactive mill tailings sites that limit the release of radon-222 from uranium byproduct materials and radon-220 from thorium byproduct materials. These engineered surface covers are designed to reduce radon emissions into the atmosphere. Increased thickness of cover material can be beneficial for reducing radon release by utilizing the increased travel time through the cover and the relative short half-life of radioactive radon.

A critical function for most engineered surface barriers is resistive or divertive in nature and designed to control surface water percolation through the cover. Reducing water contact with the wasteform helps keep the waste environmentally isolated and the contaminants contained while degradation of all barrier components is considerably slowed. Delayed or decreased contaminant transport in the saturated zone fulfills a common requirement of the engineered surface barrier. Many surface covers allow initial water infiltration to occur, however allow only a minimum to percolate out through the bottom of the cover. Covers usually consist of one or more layers so that infiltration through one layer does not equate to infiltration through all cover layers.

Surface covers employ a variety of functional mechanisms to divert water and contain waste, however most covers are of two types: conventional and evapotranspiration (ET). Conventional covers rely on the resistive properties of one or more components of the cover, such as a geomembrane, to halt downward movement of infiltration water and allow a drainage layer with high hydraulic conductivity to channel the water to the sides of the cover where it is discharged from the system. The resistive properties can be used to prevent water from reaching the waste, or radon from escaping into the atmosphere. Examples of materials with resistive properties include compacted soil or clay, sheets of synthetic material such a geomembranes, and geosynthetic clay liners. Geotextiles and geonets are often used in conjunction with granular drainage layers to provide a drainage pathway out of the cover system. Although ET removes water from conventional covers, it is the layers of low permeable material that hinder water from reaching the waste. In contrast, ET covers rely more on natural processes to remove water and would not fully function without vegetation. ET is usually the most important process removing water, sometimes removing over half of the total water input (i.e., precipitation). Water is allowed to enter the outer portion of the cover, a non-compacted soil layer with a saturated hydraulic conductivity value close to that of the surrounding environment, during periods of elevated precipitation and minimal ET. Stored water is subsequently pulled back to the surface by a network of plant roots and removed from the system by ET during drier periods. ET covers are effective when the upper soil layer has sufficient storage capacity and when sufficient evapotranspirative demand exists to remove the stored water. Engineered surface barriers may combine the features of both conventional and ET covers, e.g., installation of synthetic material below the water storage unit of an ET cover.

Components of EBS include low-permeability soil layers or compacted clay liners (CCLs), geosynthetic clay liners (GCLs), geomembranes, vegetated evapotranspiration soil layers, and drainage layers. Field studies have shown that GCLs and CCLs perform best when they are installed together with geomembranes. CCLs are generally made from varying percentages of fine-grained silts and clays. Drainage layers can be made from both natural and synthetic material and typically constructed above a resistive layer to remove fluids from the system.

EBS designs are frequently prescriptive and regulation driven. For example, RCRA Subtitle C covers have multiple layers including a biotic barrier. RCRA covers are frequently used at waste sites being remediated at under CERCLA or state-run programs and appear to work well under many circumstances. However, as is often the case, each site, or disposal site, can be associated with unique opportunities or difficulties in keeping the waste isolated. Local climate, geology, hydrology, and topography, nature of the waste and wasteform, landfill size, and other site-specific characteristics may allow alternative barrier designs. These alternative designs could require additional or special engineering components, or less engineering than prescriptive designs, depending on the uniqueness of the site and the associated analyzes.

All components of an engineered surface barrier are subject to long-term degradation and some relatively short-term degradation or initial design or emplacement flaws. Increased degradation and diminishing performance could eventually lead to failure of the engineered component as a barrier. Compacted soil or clay has been shown to change their lower as-built saturated hydraulic conductivity to conductivities closer to that of the surrounding environment, sometimes developing cracks due to desiccation, often more than an order of magnitude [9]. For ET covers, the upper storage layer may be designed for the wrong type of vegetation, develop secondary permeability features, or suffer from erosion. Drainage layers may clog due to soil infiltration, biological activity, and/or mineral precipitation [7]. Geomembranes, usually a significant component of an engineered surface barrier, can be emplaced with defective materials or seams, damaged during installation, suffer from punctures due to root penetration or undergo long-term chemical oxidation changing the polymeric properties over the long-term Depending on the disposal site's topography, climate, and waste type, erosion severe enough to reduce performance can occur. Slumping, slope failure, landslides, subsidence, and gully-forming erosion can reduce the thickness of engineered surface covers as well as increase infiltration by short-circuiting the resistive properties of a low permeable barrier or decreasing the water storage capacity of an ET cover. Climatic conditions at a cover may change during the performance period, including the frequency or magnitude of short-duration, large magnitude events. For engineered surface barriers that are a significant component of the overall EBS, features, events, and processes accelerating degradation and effecting surface cover performance need to be identified before the final cover is constructed and monitored during the postclosure and institutional control period.

1.3.2 Regulatory Compliance Criteria

During the 1950's and 1960's, the common practice of disposing of low-level radioactive waste (LLW) was disposal at sea. Commercial interest in ocean disposal declined due to public opinion and concern over costs, and ended completely by 1970. Burial at sea ended by the London Convention, "Convention n the Prevention of Marine Pollution by Dumping of Wastes and Other Matter, 1972," which has been in force since 1975. By 1971, a total of six near-surface LLW disposal facilities were licensed and operated to dispose of commercial LLW. The earliest engineered surface barriers, or covers, were earthen mounds created when trenches in which the waste was placed were backfilled using material removed during trench excavation, and subsequently compacted and graded. Due to concerns that the soil and grass covered trenches were not containing the waste material and that radionuclides were being released, the National Academy of Science (NAS) was asked to independently review near-surface disposal practices used. Although the NAS found no serious deficiencies in past federal disposal practices, it made administrative and technical recommendations to improve disposal practices. One of the recommendations made was that engineered barriers be adapted so as to work in tandem with the geological and hydrological systems.

The U.S. Nuclear Regulatory Commission (NRC) issued Part 61 of Title 10 of the Code of Federal Regulations (10 CFR Part 61) in 1982 after several years of development. The regulations covered all near-surface LLW disposals from site selection through facility design, licensing, operations, closure, and postclosure stabilization to the period when active institutional controls end. The regulation requires the use of engineered features in concert with the natural characteristics of the disposal site to contain and isolate the wastes, and included performance objectives and technical criteria. Section 3116 of the National Defense Authorization Act for Fiscal Year 2005 expanded the use of 10 CFR Part 61 as it requires NRC to monitor U.S. Department of Energy (DOE) disposal actions of waste incidental to reprocessing (WIR) of high-level radioactive waste based on the performance objectives found in Subpart C of 10 CFR Part 61.

In addition to licensing LLW near-surface disposal facilities, NRC also regulates uranium mill tailings under the Uranium Mill Tailings Radiation Control Act of 1978 (UMTRCA). This legislation addresses two types of sites: (1) those that were inactive or no longer being used when UMTRCA was passed, referred to as Title I sites; and (2) sites that were active or issued a license after UMTRCA was passed; referred to as Title II sites. Title II sites are regulated under the provisions of 10 CFR Part 40. Appendix A to 10 CFR Part 40 establishes technical and other criteria for Title II sites relating to siting, operation, groundwater protection, decontamination, decommissioning, reclamation of mills and of tailings at mill sites, and long term site surveillance. Criterion 6 of Appendix A requires an earthen cover (or approved alternative) over the tailings or waste to increase control of radioactive hazards and limit release of radon-222 from uranium and radon-220 from thorium to the atmosphere. Title I sites are regulated under the requirements of 40 CFR Part 192 and have a general license under 10 CFR 40.27.

NRC guidelines on engineered barrier performance or on monitoring disposal sites of various waste types can be found in NUREG-1388 for LLW, NUREG-1854 for WIR, NUREG-1620 for radioactive mill tailings, and NUREG-1757 for complex materials decommissioning. NUREG-1623 presents methods, guidelines, and procedures for designing erosion protection, for long term stabilization, with an emphasis on engineered surface barriers at uranium mill tailing sites in accordance with the requirements of 10 CFR Part 40, Appendix A (Title II sites) and EPA's 40 CFR part 192 (Title I sites).

1.3.3 Engineered Barrier System Performance

As previously stated, engineered surface barriers can help provide reasonable assurance that performance objectives will be met as set out in 10 CFR Part 61, Subpart C, for near-surface disposal of LLW or WIR, as well as fulfill the regulations of 10 CFR Part 40, Appendix A, for radioactive mill tailings. 10 CFR Part 61 establishes the procedures, criteria, and terms and conditions upon which NRC issues, transfers, and terminates licenses for the land disposal of radioactive LLW and covers the preoperational, operational, closure, postclosure, and institutional-control phases of the licensing process. The institutional-control requirements in Part 61.59 state that the disposal site owner must perform environmental monitoring to assure continued satisfactory disposal performance, physical surveillance to restrict access to the site, and minor custodial activities. In addition, the institutional-control period is not intended to last longer than 100 years following transfer of control; however, the determining factor on how long institutional control will remain is compliance with the performance objectives in 10 CFR Part 61.41 through 44, including the protection of the general population, protection of individuals from inadvertent intrusion and during operations, and long-term stability of the disposal site after closure so that active maintenance of the disposal site is not required. Depending on the disposal site and the waste characteristics, the institutional control period may last longer or be shorter than the suggested 100 years. NRC believes that the disposal facilities should be robust, and institutional controls should not be relied upon in order to rebuild temporary or substandard disposal components.

During a period of institutional control, the owner must perform environmental monitoring to assure continued satisfactory disposal performance, physical surveillance to restrict access to the site, and minor custodial activities. Short-term performance of the disposal site can be physically monitored with various onsite instrumentation or by remote sensing. Monitoring may detect early significant release of contaminants and may be used to verify the correctness of assumptions made and the accuracy of the results of numerical modeling, thereby reducing uncertainty. Assumptions, parameters, and features that have a large influence on the performance and/or have relatively large uncertainties should be an important part of a monitoring plan. For example, a planned cover for a waste disposal site may be an important barrier in keeping waste isolated from the environment; however, uncertainty regarding the performance of the planned cover during a potential drought-heavy rainfall cycle could not be reduced during the performance assessment process. It could not be excluded that the planned cover might fail during such conditions, either due to degraded geosynthetic clay liners or reduced evapotranspiration. A 100-year institutional control period could allow observations to be made, data to be gathered, and uncertainties to be reduced.

Monitoring is a good mechanism to manage uncertainties and to evaluate new information obtained during the institutional period. Additional information gained through various sources can reduce uncertainties and support previous predictive modeling. Monitoring is not to be used as a substitute for development of an adequate database, but rather to support the previous determination of adequacy considering uncertainty. When there is uncertainty associated with the waste disposal system, monitoring can maintain confidence in the performance demonstration. However, if the uncertainties are too large and associated with significant processes, the performance objectives will not be met and the facilities remain in the postclosure phase. In this case, a disposal site would not be allowed to enter the active institutional control period. The institutional control period should not be an opportunity to do fundamental characterization, but a time to support the compliance decision and build confidence.

Performance assessment establishes an acceptable methodology to demonstrate compliance with the performance objectives through the use of systematic risk analyses and numerical modeling. However, for periods of performance over hundreds or thousands of years, numerical models cannot be validated in a traditional sense. Assessing compliance with the Part 61, Subpart C, is expected to include adequate support of the model necessary to maintain confidence in the results of performance assessment. Model support and confidence building are important in demonstrating that, even with the inherent uncertainties, radioactive waste disposal will not endanger public health and safety. Model support should use multiple sources and various types of information such as the results of environmental monitoring, tests, information, experience with similar systems, component process models, natural and anthropogenic analogs, or independent peer reviews.

SESSION 1A:

EXPERIENCE OF THE STATES IN REGULATING FACILITIES INVOLVING ENGINEERED COVERS AND LINERS

Session Chairs:
Stephen Salomon, NRC/FSME and
Susan Jablonski, TCEQ, State of Texas

Technical Reporter:
Douglas Mandeville, NRC/FSME

2.1   Motivation

NRC Agreement States and the Navajo Nation have developed experience with the various technical and regulatory issues associated with engineered barriers at low level waste, decommissioning, and uranium mill tailings sites. The goals of this session were to: (i) discuss the regulatory activities and findings which confirm short- and long-term performance of engineered barrier systems with an emphasis on practical monitoring and modeling; (ii) identify regulations related to degradation processes that change performance; (iii) discuss monitoring devices and networks , and modeling systems; and (iv) discuss model support to gain confidence in long-term performance; and (v) identify possible evolution of regulations based on past experiences. In this session, all the State and Tribal speakers approached these questions from different perspectives based on their experiences, stage of development, and emphasized different aspects of engineered barrier performance.

2.2   Background

The NRC Agreement States represented in this session have regulatory responsibility for approval of engineered barriers at low level waste and uranium mill tailings sites. The Navajo Nation has an advisory role as a stakeholder at uranium mill tailings sites in the southwest. These groups also have experience over a range of time periods from sites currently under review or construction to sites that have been closed and are in some form of post closure monitoring. Degradation processes that impact performance include infiltration, erosion, contaminant transport, and biotic activity. Regulatory review and approval of engineered barriers is dependent on many factors; the main factors include: the regulatory framework, the waste form, site conditions, and the degradation processes previously mentioned. These factors also lead to different approaches to modeling and monitoring. Modeling programs used to support regulatory decisions include RESRAD, TOUGH2, VS2DI, PORFLOW, SWAT, SAP, PATHRAE(1-D), GWSCREEN – Version 2.5, CHILD, FLAC, FLAC3, HELP, GOLDSIM, 3DMODFLOW, MODPATH1, UNSAT-H, DUST, FOLAT, WEPP, and SIBERIA. Monitoring has been used to confirm short and long term performance of barriers. Monitoring approaches identified include: pan lysimeters, groundwater monitoring wells, time domain reflectometer probes, visual observation, and surface water sampling.

2.3   Summary of Panel Discussion

A summary of the major topics during the panel discussion follows.

- Different regulatory frameworks – 10 CFR 61 for low-level waste (LLW), 10 CFR 40 for uranium mill tailings, Agreement State regulations, and DOE Orders.
- Regulations have evolved based on past experiences
- Older sites were sometimes constructed before the regulations were in place leading to remedial action.
- There has been significant reductions in LLW generations rates which gives more time to plan for closure with covers.
- It is difficult to determine specific long-term standards.
- There is much uncertainty in relying on models for the long-term. Therefore, regulators need to be transparent and recognize these uncertainties.
- Reliable input data is a requirement for a reliable model.
- In Utah, the period of performance is evaluated quantitatively up to 10,000 years, and qualitatively, thereafter.

- For the Barnwell site in South Carolina, the performance assessment was limited to 2,000 years because of uncertainty beyond that timeframe. At the end of the institutional control period, the caps will be inspected for any necessary repairs or may need replacement.
- There could be advantages to changing LLW disposal sites more consistent with UMTRCA sites with a long term steward; dedicated funds for long-term surveillance should not be in a general fund.
- Long-term sites often must rely on DOE support because of limited resources.
- For certain areas, water balance [evapotranspiration (ET)] covers are preferable. However, design approaches and understanding may change in the future
- For the long term, proper construction and operation of covers are required to ensure compliance.
- Consideration of waste form is very important to limit ground water transport of contaminants.
- For Shiprock, seepage from the cell and impacts on the remediation system need to be taken into account. Alternatives appear to be an ET cover or moving the tailings pile. There is concern that the water is moving slowly through the pile. Both water and wind erosion may be bigger issues with the tailing being less permeable than the cover.

2.4   Summary and Recommendations

During the panel discussion, seven common themes emerged: (1) limitations to current technologies; (2) consistency; (3) cover approach; (4) flexibility; (5) improvements; (6) funding; and (7) community of practice. Further discussion and recommendations related to these themes follows.

(1) The current technologies have limitations.
Modeling and monitoring approaches are not uniform and depend upon different regulations, site conditions, waste form, and design time period. There is uncertainty in identifying appropriate design time periods; this leads to uncertainty in decision making. Communication with stakeholders is important to explain the uncertainties. Requirements can change over time based on new regulations and guidance. Federal guidance is needed early in the process, especially if depleted uranium is under consideration for disposal. Regulators have observed that containment performance for new sites appears to perform better compared to older sites, e.g., legacy sites.

(2) Consistency
Waste forms vary, which impacts the design approach of engineered barriers from site to site. In addition, degradation processes, monitoring devices and systems, codes and modeling experiences, and model support all are very site specific. The site specific nature can create difficulty in maintaining a consistent and transparent approval process. Regulators have observed different approaches at new sites compared to old sites.

(3) Cover approach
Water balance (ET) covers appear to show potential in certain arid areas such as Colorado and Washington. Engineered barrier containment design covers in both humid and arid areas (New York, South Carolina and Utah) can be enhanced by geosynthetics but appear to be limited in service life to less than 200 years assuming they are installed with impeccable quality assurance and quality control. The cover approach depends to a degree on regulatory

familiarity and comfort in addition to location. Performance monitoring post construction is essential to substantiate the analyzed condition.

(4) Flexibility
Knowledge, experience, materials, and public policy are all time dependent. This impacts both regulators and industry. What regulators believe is the right approach today may not be so in the future.

(5) Improvements
Improvements come with flexibility. Some examples of flexibility include the reduction in thickness in the radon barrier at Clive, Utah, and the enhanced caps in response to the tritium leakage at Barnwell, South Carolina. The integration of the internal containment system monitoring enhances performance.

(6) Funding
Full cost accounting of long-term maintenance and monitoring is needed. A funding source is necessary to be able to address flexibility, improvements, long-term monitoring and maintenance. The funding approach should be adapted to the particular circumstances of each site. The 100 year institutional control period for LLW may be inadequate and it may be necessary to continue some level of funding beyond the institutional control period.

(7) Community of practice
There is a large effort underway on how to improve understanding and performance of covers. At times, the effort is not coordinated and limited to addressing a need at a particular site. Meetings such as this workshop are helpful in exchanging ideas from specialists in a wide variety of disciplines and therefore should be considered as an annual event.

2.5   Extended Abstracts

The workshop organizing committee sent out invitations to the NRC Agreement States and Native American tribes requesting them to provide speakers to discuss their experiences with facilities involving engineered covers and liners. Specific speakers and panelists were identified (please see the workshop agenda in Appendix A) through cooperative discussions with the State regulators and Native American tribes and NRC's Division of Intergovernmental Liaison and Rulemaking staff. The following seven extended abstracts document these presentations and discussions during the session.

2.5.1 Modeling and Monitoring of Barrier Performance for the Planned Texas Low-Level Radioactive Waste Disposal Facility

Susan Jablonski, P.E., Peter Lodde, P.E., and Abel Porras, P.E.

Texas Commission on Environmental Quality (TCEQ)
Austin, Texas

Introduction

Although the State of Texas hosts several different types of radioactive waste disposal sites, including Title I and Title II uranium sites, this abstract focuses on the planned low-level radioactive waste (LLRW) disposal facility. On August 4, 2004, Waste Control Specialists LLC (WCS) submitted an initial license application to the Texas Commission on Environmental Quality (TCEQ) to authorize the development, operation, and closure of two facilities for the disposal of LLRW at a site in Andrews County, Texas. The site is located approximately 30 miles west of the City of Andrews, Texas, and five miles east of the City of Eunice, New Mexico. The planned facilities would be located just east of the Texas-New Mexico border. No other license applications were received by the TCEQ during the open 30-day period for receiving applications.

TCEQ Executive Director's staff reviewed the WCS license application under amended statutes and new rules, passed in 2003 and 2004 respectively. The technical review of the application included site characterization data, information on proposed operations, radiological and non-radiological impact assessments, and site closure discussions and proposed financial assurance. On August 11, 2009, a preliminary recommendation for license issuance was filed by the Executive Director, along with a draft radioactive material license and accompanying licensing order and an environmental analysis documenting the license application review were all opened for public comment.

On September 10, 2009, following a condemnation proceeding involving remaining mineral rights, the TCEQ signed and issued a license for LLRW disposal to WCS. Once constructed and operational, the Texas LLRW Comprehensive Disposal Facility will consist of two separate facilities with three planned disposal units: the Federal Waste Facility Disposal Facility - Containerized Disposal Unit (FWF-CDU); the Federal Waste Facility Disposal Facility - Non-Containerized Disposal Facility (FWF-NCDU); and the Compact Waste Facility Disposal Facility (CWF) Unit.

While each of these disposal units utilizes a barrier performance system, the engineered components of each system and their configuration differ between individual units. Each unit has a multi-layered engineered cover, and a liner with both geosynthetic and clay components. Each of the units will have a leachate collection system and systems for monitoring the groundwater in both the unsaturated and saturated porous media immediately outside of the unit barrier systems. The waste disposed in the FWF-CDU and CWF unit will be contained within packages emplaced and grouted in statutorily-required concrete canisters configured in stacks and separated by granular fill materials. In addition, these disposal units will have an outer reinforced concrete barrier or liner envelope constructed of shotcrete containing both fiber and steel reinforcement. The FWF-CDU and FWF-NCDU are authorized for the disposal of federal mixed LLRW through an overlying RCRA, Subtitle C permit that requires the addition of a leak detection system. The FWF-NCDU will not contain an outer concrete barrier/liner, but will include a leak detection system below the leachate collection system, whereas the CWF unit will not contain a leak detection system, as it is prohibited from disposal of mixed LLRW waste. Only the FWF-CDU will have all three system components: an outer concrete barrier/liner; a leachate collection system; and a leak detection system.

Given the current pre-construction and pre-operational phase of the Texas LLRW Disposal Facility, there has been a heavy reliance on modeling and engineering analyses to help predict performance of the site as authorized by the recently-issued license. The license application, and subsequent regulatory review of the planned LLRW facility, was supported by a variety of computer codes, engineering analyses, and related verifications, simulating both the short-term and long-term performance of these disposal unit barrier systems as well as by a RESRAD-based Performance Assessment simulating doses to the general public and an inadvertent intruder over a period of analysis of 50,000 years.

At the time of the submission of the applicant's Performance Assessment, the TCEQ was cognizant of several limitations in the RESRAD code relative to the assessment problem of interest. For example, the limited assessment period inherent in RESRAD necessitated a splicing together of RESRAD simulations to span the 50,000 period of analysis. A variable time-history of local erosion rates, estimated using the SWAT (Soil Water and Transport) code, could not be exercised in the performance assessment because RESRAD accepts only a constant erosion rate over the simulated assessment period. Also, the LLRW site has significant geometrical and hydrogeological complexities that cannot not be modeled with RESRAD, other than through the use of greatly oversimplified conceptual models and an inflated conservatism focused on only certain input parameters. However, because in the initial phases of the application review the applicant had selected RESRAD as their assessment tool, the TCEQ, in order to facilitate technical communication and technical transfer with the applicant, elected to also complete a performance assessment using RESRAD. In order to address shortcomings and to address support for performance objectives, a new, more advanced, and flexible performance assessment is required to be submitted for review as a license condition prior to the disposal of any waste.

Although there are projected inventories of waste to be disposed in the LLRW units that are incorporated as authorized acceptable waste, there is discussion and consideration of the potential impacts of additional volumes and radioactivity of similar waste and new LLRW waste streams. Recent discussions of the possible inclusion for acceptance of large quantities of depleted uranium, new waste streams of processed wastes, and acceptance of multiple large components from decommissioned facilities have necessitated initiation of a plan to update and reconstruct a Performance Assessment Model on platforms other than RESRAD. These new platforms need the flexibility to consider the disposal of new waste streams and the capability of incorporating, in a more realistic manner, surface geomorphological and subsurface hydrogeologic complexities present at the planned LLRW site. Additionally, factors such as the Texas requirements for modeling to peak dose and an expanded consideration of the period of analyses have also required more flexibility in the assessment modeling platform to incorporate multiple inputs from various models, evaluation over longer time periods, and capable management of inherent and propagated uncertainties.

Summary of Barrier Modeling using FLAC

Each individual engineered component of the disposal unit (concrete canister, shotcrete barrier/liner, cover system, liner system) was first designed as a separate, isolated component using elementary principals and practices from concrete design, structural mechanics, and geomechanics. A more refined analysis was conducted for the "critical" canister (i.e., one located at the bottom of a canister stack). The finite element analysis structural mechanics code SAP (Structural Analysis Program) was used to compute the three-dimensional displacement, strain, and stress fields in this critical canister and its concrete cap. The simulations incorporated the effects of thermally-induced creep in the canister materials. The initial design for the shotcrete barrier/liner was given a similar analysis using the SAP code. On the basis of the SAP analyses, appropriate modifications were made to the initial elementary designs of these barrier components.

A two-dimensional version of the continuum mechanics finite difference code, FLAC (Fast Lagrangian Analysis of Continua), was used to simulate both the static and dynamic (seismic) soil-structure interaction of all the engineered components (including granular fill materials) assembled in a total disposal unit configuration embedded within the surrounding in-situ geologic materials containing a shallow water table. The mechanical behaviors of each of the individual system components were characterized by two-dimensional, non-linear, elasto-plastic, constitutive models developed from literature compilations and laboratory testing of material samples over multiple stress paths, including dynamic testing using resonant column devices.

A large finite-difference grid (over 50,000 nodes) was able to capture the relevant spatial variation in the mechanical properties of the geologic and engineered materials. The FLAC code was capable of simulating sliding between material and engineered components interfaces. The displacements, stresses, and strains throughout the computational grid (including all barrier components and in-situ natural materials) were computed for different operational configurations of the disposal units (excavation partially open, partially filled with waste, completely filled with waste and cover in place, etc.). The effects of the excavation of a nearby disposal unit on another unit in an operational configuration were also simulated. Two different design earthquakes (applied over the lower boundary of the computational grid) were used to study the dynamic interactions of barrier system components within each completed disposal unit.

The entire set of static FLAC simulations indicated that no significant deformations (including excessive settling of the engineered covers with concomitant ponding) or stresses would occur once the engineered components were assembled into a functioning disposal facility and that disposal unit excavations would not significantly affect the nearby water table. The simulations indicated that no slope stability failures would occur during excavation or operational phases. No excessive rotations of stacked canisters were indicated so that canister tipping during an operational phase was not suggested.

The dynamic FLAC analyses indicated that stresses, strains and displacements during an earthquake would be insignificant although these computations suggested that a thickening of the shotcrete liner near engineered joints would prevent cracking due to seismically-induced small amplitude vibrations. All of these simulations were used to study the short-term performance of the barrier components. Within this modeling context, "short-term" denotes that period of time in which the units were in an operational configuration or during the time immediately after closure before the natural materials and engineered components have had time to destabilize due to erosion, corrosion, or other degrading processes.

The FLAC code was also used to model a slightly longer-term behavior of the disposal units. The effects on the soil-structure interaction of the assembled system of natural and barrier components due to the degradation of the disposal canisters, caused by excessive weathering, chemical attack or poor quality control during their fabrication, was simulated. These FLAC computations incorporated both coherent and random patterns of canister failure in order to induce significant localized subsurface volume changes within the disposal units. The FLAC simulations indicated that such volume changes, over the time frames needed for them to manifest (hundreds of years), would not result in excessive deformation or cracking in the overlying cover materials nor would cover slope reversals lead to ponding.

A three-dimensional version of the FLAC code, FLAC3D was utilized to simulate the long-term structural stability of the CWF disposal unit when used for the disposal of large components (i.e., reactor pressure vessels, steam generators, coolant pumps) from decommissioned facilities. The disposal plan intends to maintain the modular disposal framework established within the disposal unit (concrete canisters stacked in tiers) with each large component, encased in an appropriate volume fill material, substituted for an integer number of canisters within the framework. FLAC3D simulations indicated that initial plans to use a granular porous media as the encasing fill materials for the large components would induce excessive deformations in the performance cover.

Currently FLAC3D is being utilized to simulate a similar disposal plan with low-strength concrete grout serving as encasement for the disposed large components. While the results of these simulations are only preliminary these simulations indicate that the use of grout, rather than a granular media such as sand, may facilitate long-term structural stability of the performance cover. The FLAC3D computations were supplemented via the use of the SAP program. SAP was utilized to estimate stresses induced in the concrete liner via the emplacement of large components on the floor of the CWF unit. SAP was also utilized to study the stress distributions in the grout structures encasing the large components while serving as structural elements within the modular framework.

Additional FLAC3D and engineering analyses will be required to complete the analysis of the short-term and long-term structural stability of a disposal unit serving as the final disposition for multiple large components. FLAC3D will also be used to simulate the stability of facility roadways and disposal unit barrier structures during operational periods in which the large components, on transport vehicles and accompanied by appropriate lifting devices, will approach and enter the disposal unit for emplacement in a selected location within the modular framework and for grout encasement. The extreme weights of these large components (approaching 900 tons) necessitate a careful analysis of activities related to their disposal.

The described studies indicated that the FLAC code provides a valuable means to study the static and dynamic soil-structure interaction of barrier systems over short to medium periods of time. It is noted that the FLAC code was not utilized to simulate the longer term degradation of the cover materials due to geomorphic processes (erosion).

Supporting Modeling of Barrier Performance

A multi-component cover system was designed to help mitigate the effects of infiltration, erosion, intrusion, and radon exhalation. This cover system is comprised of evapotranspiration layers, a performance cover, and a bio-barrier. The short-term hydrogeologic performance of the disposal unit cover systems was studied with both the lumped-parameter code HELP (Hydraulic Evaluation of Landfill Performance) and the two-dimensional finite-difference code VS2DI (Variably Saturated Two-Dimensional Infiltration). Staff recognized the limitations of HELP in modeling infiltration and conducted their own infiltration modeling. A sensitivity analysis of the cover using HELP was performed to determine which parameters were most affected by slight changes. Then, for a variety of climatic conditions, layer properties, and boundary conditions, the distribution of possible infiltration rates into each covered disposal unit was studied. This distribution was used in the Performance Assessment completed using RESRAD by exercising the probabilistic simulation capabilities of that code. In addition, staff developed license conditions requiring more rigorous modeling to arrive at a more realistic infiltration rate using VS2DI. These license conditions require sensitivity analyses of the parameters, scenarios dealing with degradation of the layers and increased infiltration, and more appropriate boundary conditions. Staff has also independently used VS2DI in determining a more appropriate infiltration rate but has failed to come to a consensus with the applicant on the various issues concerning its implementation.

Of major concern is the possibility of the presence of saturated conditions, either now or in the period of analysis, located on the periphery of the disposal facility. The concern is that these saturated conditions may be recharged more directly and effectively at some distance away from the disposal facility. The resulting rise in subsurface water elevation could then intruce into the more permeable drainage layer, thereby bypassing a significant thickness of cover. This presents an interesting question on planning for and/or mitigating the increased saturation within the cover that can best be analyzed using a two dimensional finite element infiltration code, like VS2DI.

Since these studies have been completed, concerns have also been raised regarding potential pedogenic degradation in the hydrogeologic properties of engineered cover systems that can occur over relatively short periods of time. It is anticipated that planned performance assessments considering the possible disposal of large quantities of depleted uranium will consider the possibility of such effects. To affect these planned assessments, a model of the mass transport at LLRW disposal site based on the GOLDSIM dynamic simulation code and an extension to that code, the Contaminant Transport Module, is under development. The construction of this model will be informed by results from the three-dimensional finite-element multi-phase flow and mass transport code PORFLOW. The MAPLE computer software will also be used to provide additional supporting numerical and mathematical analyses.

Modeling of Longer Term Barrier Performance

Longer term performance of the erosional resistance of the barrier cover systems was studied using the code SWAT (Soil Water and Transport). Relative to the domain of the FLAC modeling, the SWAT modeling was regional. The region surrounding the three disposal units was divided into sub-basins (one of which contained the cover systems). Each of these sub-basins was assigned an elevation and soil properties descriptive of its surficial geologic materials. Utilizing the results of a stochastic computer code generating climatic conditions over the prescribed period of analysis, the SWAT code estimated the time history of erosion within each sub-basin (as it interacted hydrologically with all the other sub-basins). Because of the many uncertainties involved in modeling erosion over a long period of time, multiple sets of parameters were incorporated into the SWAT modeling. The SWAT results indicated the possibility that the barrier cover materials (and contiguous geologic media) might be significantly eroded over longer time frames. The simulated erosional histories were used as input to the performance assessment modeling using the RESRAD computer code. The RESRAD simulations indicated a possibility that long-term erosion could impact buried waste within the extended period of analysis and lead to large doses due to receptor exposure to external radiation. Site-specific erosion rate studies could inform and help to greatly narrow the uncertainty of impacts due to erosion over long periods. As noted previously, the RESRAD simulations assumed conventional inventories of low-level radioactive waste and did not consider the in-growth of decay products should depleted uranium be disposed of at the LLRW facility.

These analyses indicated that there are no available codes specifically designed for studying the long-term spatial distribution of erosion. SWAT and more elementary approaches based on the Universal Soil Loss Equation are not designed for the application herein described.

Monitoring of Barrier Performance

To date, the hydrogeologic conceptual model of the planned disposal site is under re-development based on prescribed additional studies. Three-dimensional transient modeling incorporating both saturated and unsaturated conditions over an area incorporating the facility boundary down to a depth of 225 feet was performed using the code TOUGH2 (Transport of Unsaturated Groundwater and Heat). This modeling incorporated laboratory derived estimates of the matric potential versus degree of saturation curves (without hysteresis) developed from site core materials. This complex computational analysis indicated that certain components of the original hydrogeologic conceptual model required additional supporting data and verification prior to any refinement of modeling on the planned disposal facility.

Pressure transducers or heat dissipation sensors are prescribed to be used at the site to measure moisture content in the engineered cover and beneath the liner of the facilities. The sensors will be linked to data loggers that will continuously relay data. It is anticipated that the moisture content profiles in the cover can be used to quantify any infiltration through the cover. This information can then be used to validate or adjust the VS2DI infiltration models accordingly.

Furthermore, the sensors beneath the liner can be used to determine whether saturated conditions are approaching the disposal unit. And, in conjunction with the leachate collection system and a leak detection system, the sensors can provide information on the movement of groundwater and can be used as early indicators of contamination and/or breech in the disposal unit. When this information is integrated with the information from the cover, a mass balance can be derived postulating the time history of the moisture content within the disposal unit.

While monitoring plans incorporating the use of both monitoring wells and lysimeters is going forward, the long-term performance of these plans is still under review as there is remaining uncertainties in the site hydrogeologic conceptual model with respect both to surficial (perched) groundwater and with respect to deeper hydrogeologic conditions. Neutron probe and electrical impedance wire systems will be used to monitor moisture content below the liner system. Installation of neutron probe access tubes must ensure no gaps between the outside of the tubes and the surrounding materials that might result in preferential flow. Additionally, there must be consideration of both the construction material of the neutron access probe access tubes and the type and strength of the neutron source used for the measurement. Neutron logging is to be manually conducted in horizontal tubes with the establishment of baseline conditions prior to facility construction and prior to waste disposal operations.

To evaluate results from FLAC, strain gages are being proposed to monitor for deformation in the canisters and concrete liners. However, it is unclear whether these gages will remain effective and/or durable over the operation life of the facility. In the long term, settlement markers will be used to monitor for erosion and degradation of the cover, canisters, and liners. Whether these devices will be successful in detecting these failure mechanisms is unknown in this stage of the project.

Conclusions

To date, there has been a heavy reliance on modeling and engineering analyses to help predict performance of the site as authorized by the initial LLRW disposal license. The license application that was the basis of the initial license issuance was supported by a variety of computer codes, some with identified shortcomings for application to this problem. Texas rule requires demonstration that acceptable dose to the general public be evaluated to peak dose, or a minimum of 1,000 years. A new, more advanced, and flexible performance assessment is required to be submitted for review as a license condition prior to the disposal of any waste at the planned facility. The TCEQ-issued license requires further work, through monitoring and modeling, to verify and demonstrate projected site performance before waste is accepted for disposal.

References

Bonaparte, R., Daniel, D.E., and Koerner, R.M. (2002). "*Assessment and Recommendations for Optimal Performance of Waste Containment Systems*," U.S. Environmental Protection Agency, National Risk Management Research Laboratory, Cincinnat, OH, EPA/600/R-02/099.

Bowders, J.J., Daniel, D.E., Wellington, J., and Houssidas, V. (1997). "Managing Desiccation Cracking in Compacted Clay Liners Beneath Geomembranes," *Geosynthetics '97 Conference Proceedings*, IFAI, Vol. 7, pp.527-540.

Design Guidelines for the Use of Fiber-Reinforced Shotcrete in Ground Support", F. Papworth, *Shotcrete*, Spring, 2002; "Quality and Shotcrete" E. Brennan, *Shotcrete,* Winter, 2005

Reedy, R. C.; Scanlon, B. R. Soil water content monitoring using electromagnetic induction, J. Geotech. Geoenvironmental Engineering 2003, Nov 2003; 10.1061/ASCE1090-02412003129:111028, 1028-1039.

Scanlon, B. R., Christman, M., Reedy, R. C., Porro, I., Simunek, J., and Flerchinger, G. N., 2002, Intercode comparisons for simulating water balance of surficial sediments in semiarid regions: Water Resources Research, v. 38, p. 1323–1339.

Scanlon, B. R., R. C. Reedy, K. E. Keese, and S. F. Dwyer, 2005, Evaluation of evapotranspirative covers for waste containment in arid and semiarid regions in the southwestern USA: Vadose Zone Journal v. 4, p. 55-71.

Texas Commission on Environmental Quality Draft Environmental and Safety Analysis of a Proposed Low-Level Radioactive Waste Disposal Facility in Andrews County, Texas. August 2008. (The Environmental and Safety Analysis is available the TCEQ website at: http://www.tceq.state.tx.us/assets/public/permitting/rad/wcs/final_draft_ea.pdf)

Texas Commission on Environmental Quality Radioactive Material License R04100. September 2009. (The radioactive material license is available the TCEQ website at: http://www.tceq.state.tx.us/assets/public/permitting/rad/wcs/r04100-final-issued.pdf)

Waste Control Specialists LLC Application for License to Authorize Near-Surface Land Disposal of Low-Level Radioactive Waste. March 2007. (Available link on the TCEQ website at: http://www.tceq.state.tx.us/permitting/radmat/licensing/wcs_license_app.html/)

2.5.2   Utah Clive Low-Level Radioactive Waste Facility

Loren Morton

Utah Division of Radiation Control
(lmorton@utah.gov)

Activities and monitoring that confirm system performance at the EnergySolutions low-level radioactive waste (LLRW) disposal facility near Clive, Utah fall into two general categories:

Short Term Monitoring – short-term performance monitoring of the LLRW disposal embankments is done with two types of devices:

A. **Pan Lysimeters** –are constructed at the outside edge of the waste, under the side-slope area, at each low-level radioactive waste (LLRW) disposal cell. The number and location of the pan lysimeters were determined by negotiation, and on average each LLRW cell has about 5 – 6 pan lysimeters designed / installed[1]. Said collection pans are: 1) installed immediately below the clay liner (before any cell construction), 2) underlain by a composite clay / HDPE flexible membrane liner (FML), 3) filled with a granular porous / permeable media, and 4) gravity drained via an internal perforated pipe that directs fluids to an observation manhole (about 30-feet deep) located about 90-feet from the outside edge of the waste. Purpose of these lysimeters is to allow collection of leachate quality samples, estimate clay liner contaminant breakthrough times, and possibly shed light on the performance assessment models approved for the facility (infiltration and contaminant transport). When fluids appear in the collection manhole EnergySolutions (ES) is required to collect and analyze a pore water quality sample. Annual video logging is also performed to confirm the physical condition and drainage status of each lysimeter's pipe. After completion of all post-closure monitoring said lysimeters can be plugged and abandoned by pressure grouting (sand / cement / bentonite mixture).

B. **Cover Test Cell (CTC)** – the LLRW Cell cover system was replicated on a test pad area located away and apart from the disposal cells. Use of an off-cell location provided opportunity to perform investigations, repairs, and maintenance without potential for adverse disturbance of a disposal cell cover. Layers found in the cover system and the CTC include (in descending order):
- Riprap (18 inches thick)
- Type A Filter (6 inches thick)
- Sacrificial Soil (12 inches thick)
- Type B Filter (6 inches thick)
- Upper Clay Radon Barrier (12 inches thick, 5.0E-8 cm/sec)
- Lower Clay Radon Barrier (72 inches thick, 1.0E-6 cm/sec)

Purpose of the CTC was to: 1) measure soil moisture content of certain layers, 2) measure internal horizontal drainage from the Type B filter, and 3) confirm evaporation calculations in the performance assessment model (HELP model, evaporative zone depth). The CTC soil layers were designed and constructed with the same geometry, layering, and in manner

---

[1]   Envirocare of Utah, Inc, 2005b; Appendix G, see EC engineering drawings 9407-4, Rev. T, May 16, 2003 (LARW Cell). For pan lysimeters at other disposal cells, see EC engineering drawings: 03022-V01, Revision 0, May 27, 2003 (Class A Cell); and 08012V-001, Revision 2, July 3, 2009 (CAN Cell).

similar to actual cover deployed on the LLRW disposal cells, and was constructed in September, 2001 and instrumentation installed and calibrated by January, 2002[2]. In general, this instrumentation consisted of three (3) different devices: water content reflectometers (WCR), heat dissipation units (HDU), and temperature probes, installed at four (4) separate depths in the CTC profile, including:

- Depth A – at about the mid-point of the Sacrificial Soil layer.
- Depth B – in the Upper Radon Barrier Layer.
- Depth C – in the Lower Radon Barrier (LRB) layer, about 3.5 – 4.0 feet above the base of the profile
- Depth D – in the LRB layer, about 1.5 feet above the base of the profile

Horizontal drainage from the Type B Filter layer was collected at the outer margin of the CTC at its lower contact with the Upper Radon Barrier, facilitated by a trough backfilled with a granular media and underlain by a sheet of HDPE FML. The trough drained to a dosing siphon to measure water flux. Vertical drainage from the CTC was collected by a pan lysimeter installed underneath the Lower Radon Barrier (granular sand, washed stone, geotextile fabric, and HDPE liner) that also drained to a dosing siphon. Both dosing siphons operated a tipping bucket to measure drainage. For comparison purposes, daily precipitation and evaporation measurements were made at a nearby, on-site meteorological station operated by ES.

Monitoring data from the CTC was provided by ES as a part of a May, 2005 proposal to combine the Class A and Class A North Cells into one (1) disposal cell (CAC Cell)[3]. In June, 2005 ES provided infiltration and contaminant transport modeling to support the CAC Cell design[4]. Said modeling included a HELP model simulation where the Evaporation Zone Depth (EZD) was set to 24 inches. CTC monitoring data was also provided in the June, 2005 ES LLRW License renewal application[2]. CTC monitoring results showed[5]:

- WCR water content in multiple layers showed seasonal cycling in the first two years (2002 – 2003), followed by an abrupt loss of seasonality in 2004 and 2005,
- A similar response in HDU water content was also seen across the same time period.
- Horizontal drainage from the Type B Filter did not correlate with CTC soil moisture content variations (both WCR and HDU), or with local precipitation patterns.

As a result, improvements were added to the renewed LLRW License and State Ground Water Quality Discharge Permit (Permit), issued in January, 2008, including: 1) submittal of a corrective action plan to resolve the CTC monitoring discrepancies[6], and 2) new requirement for annual CTC monitoring report submittal (instead of once every five (5) years)[7].

In July, 2008 in an audit submittal[8], ES concluded that:

- The WCR in the Layer A (Sacrificial Soil) was functional. However, three (3) underlying WCR devices had failed in Layers B (Upper Radon Barrier), C, and D (Lower Radon Barrier); likely as the result of soil salinity[9].

---

[2] Envirocare of Utah, 2005b, Appendix N, p. 2.

[3] Envirocare of Utah, 2005a.

[4] Envirocare of Utah, 2005c.

[5] Utah Division of Radiation Control, March 3, 2006 letter, includes a February 28, 2006 URS Corporation Round 3 Interrogatory, pp. 37 – 47.

[6] EnergySolutions LLRW Radioactive Materials License No. UT 2300249, Condition 28.

[7] Utah Division of Radiation Control, 2009a, Part I.H.17.

[8] EnergySolutions 2008a

[9] EnergySolutions 2008a, Attachment C, p. 1

- The HDU sensor in Layer A (Sacrificial Soil) had failed in December, 2001 likely as a result of installation damage and/or freezing conditions[10]. However, the HDU devices in the three (3) lower layers, B, C, and D, were functional.
- After comparison with site rainfall data, some errors were identified in surface water runoff data and lateral Type B Filter drainage data (Layer A). Said errors were likely a malfunction of the tipping bucket in each dose siphon. These errors were removed from the dataset[11].
- The malfunctioning of several different instruments on CTC affirms the difficulty of measuring soil moisture storage in the harsh environmental conditions and construction placement processes.
- Soil moisture storage in Layer B (Upper Radon Barrier) changed little between 2001 – 2004. For water balance considerations, WCR data from Layer A (Sacrificial Soil) was used.
- Lysimeter drainage from the base of the CTC was greatest in the spring of 2002, possibly due to construction water added. Average water drainage averaged 0.40 mm/yr between 2002 and 2007, but may be the product of precipitation and/or construction water[12].

Later, in the first annual CTC monitoring report for calendar year 2008, ES concluded[13]:

- The CTC infiltration rate, 0.2 mm/yr, was lower than the rates used in performance assessment modeling (1.83 to 3.64 mm/yr), as was the running average infiltration, 0.4 mm/year (2002 – 2008).
- No freezing conditions have been experienced in critical layers of the CTC, in that the lowest temperatures to date have been above 1.4° and 2.3° C in the Type B Filter and Upper Radon Barrier, respectively.

In May, 2009 ES reported an event of CTC data loss, apparently due to a failure of the data-logger[14]. Diagnosis by the ES and the manufacturer, lead to the following conclusions:

- Equipment had been damaged by lightning that had struck a power pole about 0.5 mile away from the CTC on April 9, 2009.
- A chronic power drain existed that depleted the solar charged battery[15]. It was speculated this was caused by a combination of failed WCR devices and loose wires on the WCR controller bus.
- Battery drain lead to loss of all WCR, HDU and dosing siphon data for a period of April thru July, 2009, or about 4% of the total data collected since January, 2001.

In an August 5, 2009 letter ES recommended that the WCR array be disconnected as a means to prevent system battery discharge[16]. This request was later repeated as ES reported a second event of intermittent WCR data loss in mid November, 2009[17]. Apparently, this loss was again due to the chronic power drain on the system battery. To date, the Utah Division of Radiation Control (DRC) has not approved authorized disconnection of the WCR array. In the meantime, ES has added a secondary backup battery to prevent CTC data loss.

---

[10] Ibid., p. 4
[11] Ibid., pp. 7 – 8.
[12] Ibid., pp. 9 and 12.
[13] EnergySolutions 2009a, p. 2
[14] EnergySolutions 2009b
[15] EnergySolutions 2009c
[16] Ibid., p. 4
[17] EnergySolutions 2009d

C. **Discontinued Devices: Suction Lysimeters** – in the early 1990's several suction lysimeters were installed along the west side of the LARW Cell, each positioned at the base of an unsaturated sand layer (Unit 3 Sand), immediately above a clay layer (Unit 2 Clay). This monitoring was required because: 1) the Unit 3 Sand / Unit 2 clay contact continued under the LARW Cell and dipped to the west, 2) a short distance west of the LARW Cell the Unit 3 Sand was saturated. These devices were later plugged and abandoned after: 1) an extended period of monitoring provided little if any available water sample for analysis, and 2) multiple slug tests and hydraulic analysis of the shallow aquifer showed little difference in permeability between the Unit 3 Sand and Unit 2 Clay layers.

Long Term Monitoring – long term monitoring of performance of the LLRW embankments is done via two (2) types of devices:

1. **Shallow Aquifer Monitoring Wells** – each disposal cell has a series of monitoring wells installed in the shallow unconfined aquifer. Each well is completed at a distance of about 90 feet the edge of the waste at a location outside of stormwater "V" ditch and access road. Said wells completely surround the footprint of each cell in locations that are hydraulically upgradient, downgradient and crossgradient to local groundwater flow. At the five (5) disposal cells at the Clive site (LARW, Class A, Class A North [CAN], Mixed Waste, and the 11e.(2) Cell) there are more than 80 shallow monitoring wells. Horizontal groundwater velocity at the Clive site averages about 1 – 2 feet/year. Monthly groundwater elevations are measured in the shallow monitoring wells. Previous baseline groundwater quality sampling / analysis was quarterly and semi-annual, but was recently reduced to annual frequency in light of the long travel times predicted for contaminants in the vadose zone and shallow aquifer[18].

2. **Nested Monitoring Well Pairs** –paired or "nested" monitoring wells have been installed at the Clive site to measure vertical hydraulic gradient and groundwater head in both the shallow and deep confined aquifers. Initial characterization studies indicated that a positive or upward hydraulic gradient exists at Clive, resulting in upward recharge from the deep confined aquifer. On this basis, compliance groundwater monitoring for the LLRW disposal cells has focused on shallow aquifer wells. Nested well pairs have been installed to confirm that the positive upward hydraulic gradient continues to exist at Clive.

3. **Hydraulic Gradient Monitoring** – hydraulic head monitoring is used as an early warning of disposal site hydraulic performance, and gradients are examined for compliance purposes in both the horizontal and vertical directions, as outlined below:

   A. **Shallow Horizontal Gradients** - total dissolved solids (TDS) in the shallow Clive aquifer ranges between ~ 24,000 to over 53,000 mg/l, and exhibits varying spatial density between 1.018 to 1.056 gm/cc[19]. In order to accurately determine horizontal hydraulic gradients, all groundwater head measurements are converted to freshwater equivalent (FE) elevations, or FE heads.
   Head monitoring for compliance includes preparation of shallow aquifer equipotential maps based on FE head data. In combination with shallow aquifer permeability and porosity site characterization data, it has been estimated that the horizontal average linear velocity at the Clive site (site wide average) is less than 1 ft/yr[20]. Groundwater

---

[18]   Utah Division of Radiation Control, 2009b, pp. 1-4.
[19]   Envirocare of Utah 2004, p. 8 and Table 6.
[20]   Envirocare of Utah 2004, p. 7 and Table 5.

velocity used in existing performance assessment (PA) models for the site was on the order of 1 – 2 ft/yr.

In order to ensure the Clive site is operated within the bounds of the analyzed conditions in the PA models, the Division back-calculated the respective average horizontal hydraulic gradient for each disposal cell, that corresponds to said horizontal velocity. This gradient is then used as a maximum compliance limit for groundwater monitoring purposes at each cell[21]. In the event that FE head monitoring shows the average hydraulic gradient under a disposal cell exceeds its respective compliance limit, the facility is then required to perform a root cause investigation, and perform corrective actions as warranted.

Recent ES information has found an excess horizontal gradient exists near the southwest corner of the LARW Cell. Shortly after a DRC request, an ES investigation concluded that this excess gradient was caused by the combination of a construction defect in a nearby temporary stormwater culvert and a long-standing accumulation of water at this location, resulting in artificial recharge. Culvert repair was completed in September, 2009, and confirmation monitoring of hydraulic gradient is on-going at this time.

B. **Deep Vertical Gradients (Nested Wells)** – several deep wells have been installed at the Clive site, each in close proximity to a shallow monitoring well. Nested well pairs allow comparison of hydraulic heads in both the shallow and deep aquifers. Groundwater TDS in the deep confined aquifer is generally a little greater than 20,000 mg/l, and has a density ranging between 1.016 to 1.022 gm/cc[22].

In 1991 Envirocare reported an upward hydraulic gradient in the southwest part of the site near nested well pair GW-19A / GW-19B[23]. In 2000, this hydraulic gradient was now downward, apparently caused by non-contact stormwater seepage from: 1) overflow from a stormwater retention pond in the northwest corner of adjoining Section 6, and 2) stormwater accumulations at excavations inside the 11e.(2) Cell footprint[24]. This same downward hydraulic gradient was reported again in 2005, however, Envirocare speculated that the downward groundwater flow would dissipate in time[25]. The retention pond in Section 6 is commonly referred to as the "Southwest Pond", and recent discussions with ES indicate that leakage from this pond may have also contributed to the downward gradient reversal.

In 2009 DRC asked ES to study and investigate the cause of this apparent groundwater flow reversal or downward gradient[26]. In a related move, the DRC modified the Permit to require ES to: 1) maintain a neutral or upward gradient at all nested well pairs on the site[27], and 2) submit a plan and begin de-watering the shallow aquifer near well GW-19A. In January 2010, ES submitted a plan for this de-watering, which is currently under DRC review.

4. **Shallow Ground Water Quality Monitoring** – a policy decision was made, early in the site development, that although the groundwater at Clive is saline, the State would protect it as if it were a drinking water resource. This is authorized by the Utah Ground Water Quality

---

[21] For details, see Utah Division of Radiation Control 2009c, Parts I.E.25 and I.H.2(d).
[22] Envirocare of Utah 2004, p. 8 and Table 6.
[23] Bingham Environmental, p. 15 and Figure 1.10.
[24] Envirocare of Utah 2000, p. 11
[25] Envirocare of Utah 2004, pp. 7-8
[26] Utah Division of Radiation Control 2009a
[27] Utah Division of Radiation Control 2009c. Part I.E.26

Protection Rules[28], and was done to both encourage superior engineering containment of the waste (best available technology), and protect public health and the environment from waste disposal that would largely benefit waste generators outside Utah.

Groundwater quality sampling and analysis from Clive wells dates back to before 1992. Using this data, groundwater protection levels (GWPLs) have been developed that are based on either: 1) mean contaminant concentration plus two standard deviations (95% upper confidence limit), or 2) State Ground Water Quality Standards (GWQS)[29], whichever is greater[30]. The GWQS are largely derived from drinking water standards set by the U.S. Environmental Protection Agency.

Well specific GWPLs have been determined for multiple analytes, including mobile radioisotopes such as: carbon-14, iodine-129, technetium-99, tritium, and several uranium isotopes. When excess concentrations are detected, ES is required to accelerate their sampling to confirm the excess. After multiple samples provide confirmation, the company is required evaluate the cause(s) / source(s) and report on possible remediation options. Adjustments (increases and decreases) in GWPLs are occasionally approved by the DRC after determination that: 1) a lack of a trend exists in time-series groundwater quality data, 2) the data has a normal statistical distribution, and 3) any increase was not due to embankment leakage. To date, excess groundwater concentrations observed at Clive have been due to either: 1) leaching of vadose zone contaminants due to shallow groundwater mounding, or 2) recharge of contaminated stormwater to the shallow aquifer via broken or leaking pipes.

5. **Other Facilities With Potential to Effect Site Performance** – in addition to above discussion, on-site wastewater ponds have the potential to adversely effect groundwater performance by increasing horizontal hydraulic gradients, and causing downward groundwater flow. Wastewater held in these ponds is from container decontamination operations, and contact stormwater from both the LLRW and 11.e(2) operations, and is collected and disposed in one or more evaporation ponds on site. A leak detection system (LDS) under these ponds provides early warning of leakage, and is made possible by double FML membrane liner construction and a leak detection system (LDS) that gravity drains to a collection / detection sump. An inclined access pipe then allows a pump to remove accumulated leachates. Compliance limits are established for each evaporation pond using EPA guidance as based on: 1) a maximum daily average allowable pump discharge (gal / acre / day), and 2) a maximum LDS head of 1 foot, as measured from the lowest elevation of the lower FML (not including the LDS sump) by a pressure transducer fixed to the LDS pump. Early warning of leakage allows the Licensee to intervene early, before a release has impact on local groundwater gradients and quality.

Conclusions on Devices to Measure Disposal Cell Performance

Short-term monitoring devices include: pan and suction lysimeters and the CTC. Pan lysimeters continue to be functional. Suction lysimeters have been discontinued after determination they were largely ineffective.

CTC monitoring devices include WCR and HDU instruments, a drain to collect lateral Type B filter drainage, and a pan lysimeter to collect vertical drainage. CTC history shows installation, operation, and monitoring can be problematic. WCR failure appears to be rooted in corrosion

---

[28] Utah Administrative Code (UAC) R317-6-4.7.
[29] Found in Utah Ground Water Quality Protection (UGWQP) Rules, UAC R317-6.
[30] Utah GWQP Rules, UAC R317-4.7 and 6.16.

by saline soil conditions. One shallow HDU device appears to have failed due to installation damage or by freezing conditions. Nearby lightning strikes appear to have damaged sensitive electronics, leading to chronic battery discharge and partial loss of data. Vertical drainage collected to date may be the product of both precipitation and release of construction water. It is not known yet if the CTC has reached a steady-state flow condition. On a positive note temperature measurements in the CTC show freezing conditions have not reached the upper radon barrier.

Long term monitoring includes water level measurements and reporting from shallow wells to measure horizontal hydraulic gradients, and from paired or nested wells to measure vertical hydraulic gradients. Where evidence of significant change in flow directions / gradients is apparent, additional work is required to determine the cause and mitigate it as soon as possible. The goal of this monitoring is to maintain groundwater velocity and flow directions within the bounds of the analyzed conditions found in the performance assessment models.

Shallow wells are also used for ground water quality monitoring for early detection of possible seepage releases from the embankments GWPLS for multiple analytes, including mobile radioisotopes, have been determined on an individual well basis, and are based on either background concentrations or the respective State GWQS, whichever is greatest. To date, excess contaminant concentrations observed in groundwater have not been caused by seepage from the disposal embankments.

References

Bingham Environmental, "Hydrogeologic Report Envirocare Waste Disposal Facility South Clive, Utah" final consultants report prepared for Envirocare, October 9, 1991, 33 pp. with tables, figures and appendices.

EnergySolutions 2008a, "Radioactive Material License No. UT 2300249 License Condition 28: Unsaturated Flow Post Model Audit Assessment", July 7, 2008 company letter response from Dan Shrum to Dane Finerfrock, 14 pp. and Attachment B chronology (5 pp.), Attachment C – January 31, 2008 Desert Research Institute Report (12 pp.) [includes Appendix A, an October 18, 2007 Report "Calibration of Water Content Reflectometers for the EnergySolutions Site" by Craig H. Benson and Xiaodong Wang.

EnergySolutions, 2009a, "Radioactive Materials License No. UT 2300249, License Condition 28B: Cover Test Cell Annual Report – 2008", February 4, 2009 company report, 3 pp with attached CD.

EnergySolutions 2009b, "Groundwater Quality Discharge Permit No. UGW450005. Part I.E.4 Unsaturated Soil Content Monitoring, Loss of Data, May 11, 2009 company letter report from Sean McCandless to Dane Finerfrock, 2 pp.

EnergySolutions 2009c, "Follow-Up to May 11, 2009 Letter: Groundwater Quality Discharge Permit No. UGW450005, Part I.E.4 - Unsaturated Soil Content Monitoring, Loss of Data", August 5, 2009 company letter from Sean McCandless to Dane Finerfrock, 11 pp.

EnergySolutions 2009d, "Groundwater Quality Discharge Permit No. UGW450005: Amendment Request to Part I.E.4 and Request to Add Appendix D - Cover Test Cell Monitoring, Cover Test Cell System Modification Notification", December 28, 2009 company letter from Sean McCandless to Dane Finerfrock, 3 pp., includes attachment.

Envirocare of Utah 2000, "Revised Hydrogeologic Report for the Envirocare Waste Disposal Facility Clive, Utah January 2000 Version 1.0", consultants report prepared by Pentacore Resources, 16 pp. with tables, figures, and appendix.

Envirocare of Utah 2004, "Revised Hydrogeologic Report for the Envirocare Waste Disposal Facility Clive, Utah August, 2004 Version 2.0", found in Appendix D of the June 21, 2005 Envirocare of Utah "Radioactive Materials License No. UT2300249 Revised License Renewal Application".

Envirocare of Utah 2005a, "Radioactive Material License # UT 2300249 and Ground Water Quality Discharge Permit No. UGW450005. Amendment and Modification Request - Class A Combined Embankment", May 27, 2005 submittal.

Envirocare of Utah 2005b, "Envirocare of Utah, LLC License Renewal Application Radioactive Material License Number UT 2300249 Text Revision 2 June 20, 2005", May 27, 2005 submittal.

Envirocare of Utah 2005c, "Radioactive Materials License #UT 2300249 and Ground Water Quality Discharge Permit No. UGW450005. Class A Combined Embankment - Infiltration and Transport Modeling", June 24, 2005 consultants report by Whetstone Associates.

Utah Division of Radiation Control 2006 "DRC Response to Eight Submittals by EnergySolutions Regarding Proposed Class A Combined (CAC) Disposal Cell: Request for Additional Information, Round 3 Interrogatory", March 3, 2006 letter from Loren Morton to Tye Rogers (2 pp.); includes a URS Corporation February 28, 2006 Round 3 Interrogatory, 47 pp.

Utah Division of Radiation Control 2009a, "Vertical Hydraulic Gradient at the EnergySolutions Clive Facility: DRC Summary of December 23, 2008 Meeting", January 8, 2009 agency letter from Dane Finerfrock to Daniel B. Shrum (4 pp.).

Utah Division of Radiation Control 2009b, "Ground Water Quality Discharge Permit UGW450005 Statement of Basis, Low-Level and 11e.(2) Radioactive Waste Disposal Facility, EnergySolutions LLC", agency explanatory document for proposed permit modification, September, 2009, 23 pp. and 1 attachment.

Utah Division of Radiation Control 2009c, "Ground Water Quality Discharge Permit, No. UGW450005", modified discharge permit issued on December 23, 2009, 66 pp. with appendices.

2.5.3 Overview of the Performance and Use of Engineered Barriers at the Barnwell LLRW Disposal Site

Susan E. Jenkins

Division of Waste Management
South Carolina Department of Health and Environmental Control (SC DHEC)
jenkinse@dhec.sc.gov

The Barnwell Disposal Site (Site) is located in Barnwell, South Carolina and has been operating since 1971 as a low-level radioactive waste disposal facility on state-owned land leased to Chem-Nuclear Systems (CNS) LLC (a wholly owned subsidiary of EnergySolutions). The licensed disposal area is 235 acres and 119 acres have been used for disposal and subsequently capped. The site is expected to continue operating receiving waste from members of the Atlantic Compact for a period of about 30 years (CNS, 2005).

Three main types of engineered barriers are used at the Site to improve the disposal facility's ability to meet the performance objectives established in State and Federal regulations. These barriers are the disposal vaults, engineered trenches and the enhanced cap.

The South Carolina Radioactive Materials Regulation 61-63 (State regulations) has general requirements regarding the use, maintenance and performance of engineered barriers. More specific requirements are found in the facility license and approved procedures.

State regulations require that engineered barriers be designed and constructed of materials having physical and chemical properties so as to provide reasonable assurance that the barriers will maintain their functional integrity under all foreseeable conditions for at least the institutional control period (minimum 100 years). Covers must be designed to resist degradation by surface geologic processes and biotic activity. The site operator, CNS, is required to observe, monitor, and carry out the necessary maintenance and repairs at the disposal site (including engineered barriers) until site closure is complete and the license is transferred by SC DHEC to a site custodian. Maintenance must be conducted during the post-closure care period as well. South Carolina Regulations additionally require plans for taking corrective measures if the environmental monitoring program shows migration of waste that would indicate that the performance objectives may not be met.

Engineered Trenches

Disposal of waste occurs in engineered trenches. The trenches are designed to prevent water from accumulating in the waste zone and so that the bottom elevation is at least five (5) feet above then current historic high groundwater elevations. Currently, the vast majority of trench sumps are dry after the completion of enhanced caps. (CNS, 2009a) Tritium in water samples from the sumps have been at background levels since 2007. Engineered trenches are important engineered barriers because they are designed to prevent long-term (hundreds of years) water contact with waste.

Trench bottom liners have not been required at the Site. State regulations require that site design features shall be directed toward long-term isolation and avoidance of the need for continuing active maintenance after site closure. Such a design would require an extensive capability to collect trench water for management thereby increasing long-term maintenance requirements. Currently accepted modeling predicts the Site will continue to meet performance objectives without the use of bottom liners. Additionally, the installation of bottom liners would contribute very little in mitigating the existing plume which data indicates originates from the source term from older trenches. (CNS, 2009b)

Engineered Disposal Vaults

The use of engineered disposal vaults for Class B and C waste began in 1989 due to concerns over mechanical creep reported in high density polyethylene high integrity containers (HICs). (DHEC, 1989) State regulations required the use of engineered barriers for all waste classifications beginning in 1995 to further increase the stability of the disposal site. As a condition of the license, the Site uses concrete vaults to meet this requirement.

Disposal vaults also provide for intruder protection. The alignment and space between vaults is relatively uniform in geometry thereby improving the effectiveness of flowable fill and further reducing void spaces (which provide a preferential pathway for contaminants and which may lead to settling.) The vaults are therefore expected to reduce the incidence of settlement and subsidence.

Engineered Enhanced Caps

The installation of enhanced caps to cover existing clay caps at the Site was in response to monitoring well results which showed tritium in offsite wells. Above background levels of tritium were measured in monitoring wells on site in 1978 and in monitoring wells outside the site boundary in 1991 (CNS, 2010). Tritium was first detected in Mary's Branch creek (located about 2800 feet southwest of the southernmost trenches) in 1991 (CNS, 2010). It was hypothesized that if water could be prevented from percolating through the waste, tritium migration would be reduced over time and compliance with performance objectives for dose to an offsite member of the public would be maintained.

The integrity of the caps are monitored, through routine inspections The caps are inspected for the presence of unwanted vegetation such as trees and shrubs, for depressed areas caused by erosion, settling, etc., for the condition of the vegetative cover. Regular maintenance assures continued performance and repairs are made as necessary while allowing for sufficient periods of observation following initial discovery. Several repairs have been made on the older caps due to settling.

Performance

Current data indicates the performance of the disposal site and its engineered barriers is acceptable. Hypothetical dose from air emissions and direct radiation are at background levels[31] (CNS, 2003 and CNS, 2009c). Current measurements show the hypothetical dose rate at the compliance location through the groundwater pathway is less than 5 mrem per year total effective dose equivalent (TEDE) (CNS, 2009b). An environmental radiological performance verification projecting measured radionuclide concentration through the groundwater pathway

---

[31] Since Barnwell is an operating disposal site, there is radiation dose from direct radiation and air emissions due to waste disposal operations. However, the radiation dose from air emissions and direct gamma radiation are expected to be negligible after the waste disposal is completed.

showed the maximum hypothetical annual dose is approximately 13 mrem per year TEDE (CNS, 2003). Currently, a new groundwater model is being developed which will provide an updated assessment of the site performance related to offsite dose at the creek.

References

References may be obtained by e-mailing Susan Jenkins: jenkinse@dhec.sc.gov

South Carolina Radioactive Materials Regulation 61-63, (available at SC DHEC's website www.scdhec.gov and via direct link http://www.scdhec.gov/health/radhlth/61-63.htm)

CNS, 2010, *Work Plan Addressing Performance Objective E, Groundwater Flow Model and Environmental Radiological Performance Verification*, CNS Report #BEDL-10-005.

CNS, 2009a, *Performance Objective O, Report Addressing Trench Water Management*, CNS Report #BEDL-09-033.

CNS, 2009b, *Performance Objective N, 2009 Annual Trending Data Report*, CNS Report #BEDL-09-042.

CNS, 2009c, *Performance Objective D, Direct Gamma Radiation from Buried Waste*, CNS Report #BEDL-09-019.

CNS, 2005, *Interim Site Stabilization and Closure Plan for the Barnwell Low-Level Radioactive Waste Disposal Facility, 2005 Closure Plan, June 2005*, CNS Plan #PL-CNS-05-001.

CNS, 2003, *Environmental Radiological Performance Verification of the Barnwell Waste Disposal Facility Summary, July 2003*, CNS Report #BEDL-03-003.

DHEC, 1989, *Letter, H. G. Shealy, BRH to M.T. Ryan, CNSI dated 01/9/89)*

2.5.4 Washington State's Experience with Decommissioning and Evaluation of Cover Designs for Low-Level Radioactive Waste and Uranium Mill Tailings Facilities

Gary Robertson

Office of Radiation Protection
Washington State Department of Health
gary.robertson@doh.wa.gov

Washington State has a long radioactive waste disposal history with the US Ecology commercial low-level radioactive waste disposal facility (USE), operating since 1965, and with the Dawn Mining and milling Company (DMC), operating since 1956. The state's experience is unique in that these two facilities were first licensed by the Atomic Energy Commission, and regulatory authority was later transferred from the federal government to Washington State's Department of Health (DOH) in 1966 for USE, and 1969 for DMC. The reclamation and closure of these two facilities is much more difficult and complex because their construction and operations predate 10 CFR Part 61 and 10 CFR Part 40 regulatory standards.

The state has also licensed two other uranium mills, the Western Nuclear, Incorporated (WNI) uranium mill facility in 1979, a conventional mill, and the Joy Mining Company (Joy) uranium mill in 1985, a small experimental bog extraction facility. Interestingly, these two newer facilities are now reclaimed and closed, whereas the older facilities are not. The WNI uranium mill was reclaimed in 1999 and successfully transferred to US DOE in 2001, and the Joy uranium mill was reclaimed to a free released standard in 1995 and returned to the Washington State Department of Natural Resources for unrestricted use.

During the WNI reclamation and closure review process in the 1990's, the state rejected the company's proposal to reclaim the site with a standard compacted clay barrier. What made the WNI closure more challenging was that the area's climax plant community was the ponderosa pine tree. When cover performance was evaluated, the ponderosa pine root intrusion into the compacted clay barrier was determined to adversely impact long-term (1000 years) radon emanation rates.

WNI was directed to consider an alternative self-healing, thick homogenous evapotranspiration (ET) cover design. At the time there was very little regulatory flexibility for alternative covers and WNI was required to prove that the alternative could be as effective as a conventional clay barrier. WNI was able to demonstrate how a thick ET soil cover could satisfactorily store precipitation while plants are dormant, and how evapotranspiration would remove moisture during plant growing periods. WNI was also able to demonstrate how the thick cover was able to meet long-term radon emanation requirements.

The WNI closure project was able to verify that an ET cover can be designed and constructed to accommodate natural site processes, and also be able to sustain a high level of performance with little to no active maintenance. The 1000-year cover design performance criteria for stability, infiltration, and radon emanation were demonstrated through modeling and engineering calculations. WNI considered the following factors when evaluating the engineering design:
- Long-term climate change.

- Vegetation succession and responses to potential climate change.
- Worst case scenarios from fires, pests, and invasive plant communities.
- Effects of soil permeability.
- Effects from soil erosion.
- Effects of burrowing animals.
- Effects of soil development processes on water storage, permeability, and ecology.

WNI looked at and identified the long-term stable natural analogs at the site and incorporated them into the cover design. Natural analogs were also used to provide insight into how site-specific ecological processes may influence the performance of the covers.
The key takeaways or elements identified during the WNI project that are extremely important to the successful construction of a cover include:

- Early and frequent stakeholder involvement.
- Transparency.
- Having on staff seasoned technical experts that include at least a geohydrologist, geochemist, geotechnical engineer, civil engineer, soil scientist, plant biologist, seismologist, meteorologist, quality assurance engineer, and health physicist.
- Depending on the construction schedule to conduct weekly (if not daily) site visits by the state technical expert(s).
- Holding routine periodic meetings with the company and also with the NRC.
- Keeping a good working relationship with the design contractor and the construction contractor.
- Identifying and characterizing an adequate nearby soil and rock borrow source.
- Simplifying the complexities of construction specifications.
- Establishing well defined and agreed to QA/QC requirements for the company and the regulator.
- Hiring an experienced ET cover construction contractor.
- Use of appropriate plant material.
- Use of native plant communities.
- Use of proper planting techniques.
- Adequate monitoring and surveillance program for infiltration, subsidence, erosion, and vegetation.
- Adequate financial surety for long-term stewardship.
  - 100 years for commercial low-level radioactive waste disposal facilities.
  - Perpetuity for uranium mill disposal facilities.

With the successful completion and transfer of the WNI site to USDOE, DOH decided to use the same closure design methodology at DMC and USE. However it is important to note that each design differed and is specifically tailored to individual site characteristics and properties such as climate, disposal inventory, and regulatory requirements.

In the case of US Ecology, several closure plans had been submitted, the first of which was in 1983. Subsequent closure plans were submitted in 1987 and 1990. In each case, DOH required amendments to the plan. The most recent closure plan was submitted in 1996, and was approved. This plan addressed surety funding, cover design, cover schedule, institutional controls, and long-term environmental monitoring (100 years). The approved closure plan also demonstrated that hypothetical post-closure doses will be no higher than 25 millrem per year to any person living adjacent to the site.

DOH assumed that all covers at USE would catastrophically fail at 500 years and that infiltration through the covers would return to natural rates. The assumptions on cover reliability depend, in part, on the incorporation of the 1000-year cover design stability criteria in the Uranium Mill Tailings Recovery Act and on the ability of the state or other caretaker to maintain the cover if it should be damaged during the first 100 years following closure. Performance criteria for the cover include:

- Water infiltration rate through the cover at less than or equal to 0.5 mm/year.
- Radon 222 emanation rate through the cover at less than or equal to 0.62 pCi/m²s.
- Cover depth equal to or greater than five meters.
- Offsite resident dose less than or equal to 22 millirem per year.
- Onsite resident dose less than or equal to 107 millirem per year.
- Compliance with Minimum Technical Requirements for RCRA Landfills as defined in RCRA guidance document--Landfill Design Liner Systems and Final Cover, EPA PB 87-157 657/AS, 1987.

The Radiological Risk Assessment for the 2004 Environmental Impact Statement for closure of the US Ecology low-level radioactive waste disposal facility posed additional challenges because of the variety of isotopes that must be evaluated.

The source term used for the Radiological Risk Assessment includes all radioactive waste disposed at the site. Future source term projections were based on waste activity from 1993 through 1996, plus the Trojan and Washington Public Power Supply reactor vessels.

The total commercial LLRW site inventory contains about 622 separate isotopes. A majority of these radionuclides are short-lived or of minimal activity. Screening tools were used to identify the radionuclides with the highest likelihood of contributing to dose. Twenty-one radionuclides passed the initial screening criteria of a half-life greater than 5.5 years and a total activity of at least 1 curie. The existing source term for the 21 radionuclides is 1.1 million curies. Future source term (2002 through 2056) for these radionuclides was estimated to be 4,490 curies.

Cover infiltrating water is the primary mechanism of radionuclide transport to groundwater. UNSAT-H, a numeric model, was used to predict infiltration through the cover designs. In UNSAT-H, the cover characteristics most important for controlling infiltration are percent gravel, percent silt, and depth of the upper silt loam layer. Only the top layers of the covers, down to, but not including the low-permeability barriers, were included in the UNSAT-H modeling. The barrier was not included in the model because it was considered secondary to the evaporative properties of the cover in controlling infiltration. The evaporative properties of the covers were assumed to fail at 500 years. Following failure, infiltration rates of all cover designs were assumed to return to a background infiltration of 5 millimeters per year.

The original groundwater modeling done in the Draft EIS used the GWSCREEN Version 2.5 code. For the Final EIS, the modeling was revised to reconcile measured concentrations of radionuclides in the unsaturated zone with model-predicted values. Evaluation of concentrations in the unsaturated zone required a new conceptual and mathematical model of waste disposal and radionuclide transport in the unsaturated zone. The new model, the Disposal Unit Source Term Model (DUST) (Sullivan 1996) and the First Order Leach and Transport (FOLAT) more accurately reflects the waste disposal history, time-variable infiltration, and radionuclide transport in the unsaturated zone. Radionuclides used to recalibrate transport through the vadose zone include Ni-63, Sr-90, Tc-99, Pu-239/240, and U-238. The recalibration determined there was a small but highly mobile fraction of waste, which has moved essentially at the rate of water through the vadose zone.

Of the more than 600 radionuclides disposed at the LLRW disposal site, very few have a long enough half-life, large enough source term, and are soluble enough to cause a potential impact to groundwater. The initial screening identified 15 radionuclides as potentially important in terms of their impact to groundwater: C-14, Cl-36, H-3, I-129, Pu-238, Pu-239, Pu-240, Pu-242, Ra-226, Tc-99, Th-230, Th-232, U-234, U-235, and U-238.

Of the 15 radionuclides, the modeling predicted that seven would be important for groundwater during the zero to 10,000-year time period. These seven radionuclides are I-129, Tc-99, U-238, H-3, C-14, and the mobile fractions (MF) of U-234 and Pu-239. The contributions from Ni-63 and Sr-90, although in measurable quantities in the vadose zone, were not significant in terms of future groundwater concentrations. The lack of influence of Ni-63 and Sr-90 was due to the short half-lives (100 and 30 years respectively) of these two nuclides.

Direct intrusion into the waste is one of three pathways evaluated in the Radiological Risk Assessment. The probability of a person, animal, or plant coming into direct contact with the waste is affected both by the cover thickness and by the materials in the cover. The covers will be at least five meters thick and is expected to be effective at preventing direct contact by an inadvertent intruder.

Biotic intrusion includes both plant roots and burrowing animals. USDOE summarized the published information on plant rooting and animal burrowing depths for Hanford. The deepest burrowing animal was the harvester ant at 8.9 feet, and the badger was the deepest burrowing mammal at 8.2 feet. The plant species with the greatest average maximum rooting depth is antelope bitterbrush at 9.7 feet. Based upon this information, the direct contact exposure pathway of plants or animals is considered negligible for all cover designs.

Gas emanation through the cover designs was modeled for 10,000 years post-closure. Radon, C-14, and H-3 are the three potential contributors to dose. Radon is the most significant contributor to dose. NARM is the primary contributor of radon. The Radiological Risk Assessment projects the highest radon concentrations to be indoors. One driving assumption for the indoor radon dose is that a resident intruder will build a home with a basement. Building requirements for access and egress from a basement dictate that a seven-foot excavation depth is reasonable for new construction.

Depth to waste, the presence of clay, the type of low-permeability barrier, and the soil moisture content are three main factors that affect the projected radon flux. The modeling has shown that radon is significantly reduced by burying future waste containing radium 226 at a depth of 23 feet below grade.

Groundwater contamination has the potential to impact the greatest number of individuals. The primary route for exposure to individuals is direct ingestion of groundwater used as drinking water. Other pathways for groundwater include exposure via inhalation while in steam rooms (as is the case for the Native American sweat lodge), consumption of plant and animal products that have been irrigated with contaminated water, and external exposure such as bathing.

The Radiological Risk Assessment includes an uncertainty analysis for the predicted hypothetical doses. The analysis only evaluates the GeoSynthetic Cover and the Rural Resident Adult Scenario. The purpose of the uncertainty analysis is: (1) to determine how accurate the single-point dose projections are in determining the Maximally Exposed Individual; and (2) to calculate the most probable dose estimate. The uncertainty analysis considers the possible range of a given parameter such as drinking water ingestion rate, amount of food grown, and time of residence. Ranges were determined for selected parameters, and a Monte Carlo approach was used to determine uncertainty. This approach allows each parameter to vary within a predicted distribution in order to determine the most likely dose and the 95th percentile dose.

The uncertainty analysis evaluated uncertainty for the offsite and onsite resident at 60 years, 1000 years, and 10,000 years after closure. The most probable dose is represented by the mode value, and the 95 percentile represents the maximally exposed individual. The results of the uncertainty analysis indicate that the maximum single-point dose projections presented are likely to represent the MEI during the first 500 years after closure, but they under-represent the MEI during the 500 to 10,000-year period. The uncertainty analysis shows that the single-point dose estimates are less certain after 1000 years. The mode value represents the most probable dose an individual would receive.

Additional mitigation measures that are being implemented at the USE include:
- Evaluate ways to reduce the offsite materials needed for cover construction.
- Institutional controls for the foreseeable future.
- License limits for radionuclides predicted to contribute to a post-closure dose; Ra-226, H-3, I-129, Tc-99, U-238, C-14, U-234, and Pu-239.
- Deeper burial of discrete NARM.
- Secondary containment for radionuclides predicted to contribute to groundwater concentrations; H-3, I-129, Tc-99, U-238, C-14, U-234, and Pu-239.
- Continued environmental monitoring to refine future hypothetical groundwater concentrations.
- Establish and maintain vegetation on the completed cover.
- Use of the Close-As-You-Go Schedule for future trenches.

There is an interesting and somewhat questionable inconsistency between the institutional control periods for low-level radioactive waste facilities (100-year institutional control) and uranium mill facilities (perpetual control and maintenance). Institutional knowledge and experience at closed or abandoned uranium facilities has illustrated the importance of not only relying on engineered barriers, but also incorporating long-term institutional controls for as long as the waste presents a hazard to the public and/or the environment. Experience has shown that ensuring successful long-term closure depends on three interrelated controls that include site selection, engineered barriers, and long-term institutional controls. Unfortunately, when it comes to low-level waste regulations, we are only allowed to consider a 100-year institutional control period, following final facility closure.

Current studies have demonstrated how quickly chemical, physical, and biological processes begin working once a site closure and how they can negatively influence overall site performance. It is extremely important to establish an active, robust, and sustainable long-term stewardship program to counter these processes. For example, continued long-term monitoring and inspection are necessary at closed uranium mill facilities as well as low-level waste facilities in order to identify and institute early intervention to ensure long-term protection of engineered barriers from failure caused by erosion, bio-intrusion, subsidence, barrier failure, infiltration, and contaminate seepage. The same is true for long-term site-specific environmental sampling that allows for sample flexibility which is based on changing site conditions. The long-term financial surety requirement found in the uranium mill regulations should also be incorporated into the low-level radioactive waste regulations to ensure adequate funding for perpetual control and maintenance at these sites.

References

1.	EPA, 2000a, National Drinking Water Regulations, Radionuclides, United States Environmental Protection Agency, 65 F.R. 39650, Final Rule, 40 CFR Part 141, December 2000.

2.	DOH, 1998b, Evaluation of Engineering Qualities of Closure Alternatives for the Commercial Low-Level Radioactive Waste Site DEIS, Washington Department of Health, Olympia, Washington.

3.	DOH, 1999, Technical Evaluation Report for the 1996 US Ecology Site Stabilization and Closure Plan, Washington State Department of Health, Olympia, Washington.

4.	DOH, 2002, Summary of US Ecology Site Investigation Results for Radionuclides, Washington Department of Health, Olympia, Washington.

5.	DOH, 2002a, Evaluation of Data at the Commercial Low-Level Radiological Waste Site, Washington Department of Health, Olympia, Washington.

6.	Fayer, M.J. and T.J. Jones, 1990, UNSAT-H Version 2.0: Unstated Soil Water and Heat Flow Model, PNL6779, Pacific Northwest Laboratory, Richland, Washington.

7.	ICRP (International Commission of Radiological Protection), 1990, Recommendations of the International Commission on Radiological Protection, ICRP Publication 60, Oxford: Pergamon Press.

8.	INTERNATIONAL COMMISSION ON RADIOLOGICAL PROTECTION, Radiological Protection Policy for the Disposal of Radioactive Waste, ICRP Publication 77, Pergamon Press, Oxford and New York (1997).

9.	INTERNATIONAL COMMISSION ON RADIOLOGICAL PROTECTION, Radiation Protection Recommendations as Applied to the Disposal of Long Lived Solid Radioactive Waste, ICRP Publication 81, Pergamon Press, Oxford and New York (2000).

10.	INTERNATIONAL ATOMIC ENERGY AGENCY, Application of the Concepts of Exclusion, Exemption and Clearance, IAEA Safety Standards Series No. RS-G-1.7, IAEA, Vienna (2004).

11.	INTERNATIONAL ATOMIC ENERGY AGENCY, Regulatory Control of Radioactive Discharges to the Environment, Safety Standards Series No.WS-G-2.3, IAEA, Vienna (2000).

12.	INTERNATIONAL ATOMIC ENERGY AGENCY, Predisposal Management of Radioactive Waste, Including Decommissioning, Safety Standards Series No. WS-R-2, IAEA, Vienna (2000).

13.	NCRP, 1987, National Council on Radiation Protection and Measurements, Ionizing Radiation Exposure of the Population of the United States, Report No. 93, Bethesda, Maryland.

14.	NCRP, 1993, National Council on Radiation Protection and Measurements, Limitation of NRC, 1977, NUREG 0170, Final Environmental Statement on the Transportation of Radioactive Material by Air and Other Modes, Volumes I and II, U.S. Nuclear Regulatory Commission, Washington, D.C.

15.	NRC, 1982, Final EIS for 10 CFR Part 61 – Licensing Requirements for Land Disposal of Radioactive Waste, U.S. Nuclear Regulatory Commission, Washington D.C.

16. NRC, 2003, Long-Term Hazard of Millstone Unit 1's Missing Spent fuel Rods Potentially Disposed at the Hanford Commercial Low-Level Radioactive Waste Disposal Facility, U.S. Nuclear Regulatory Commission, Washington D.C., January 6, 2003, FR Volume 68, No. 3, 588-589,

17. Rood, A.S. 1999, GWSCREEN: A Semi-analytical Model for Assessment of the Ground Water Pathway from Surface or Buried Contamination, Theory and User's Manual Version 2.5, INEEL/EXT-98-00750, Rev. 1, February, Idaho National Engineering and Environmental Laboratory, Idaho Falls, Idaho.

18. Rood, A.S., 2003, Groundwater Concentrations and Drinking Water Doses with Uncertainty for the U.S. Ecology Low-Level Radioactive Waste Disposal Facility, Richland Washington, K-Spar, Inc. Scientific Consulting, Rigby, Idaho.

19. Rood, A.S., 2003a, FOLAT: A Model for Assessment of Leaching and Transport of Radionuclides in Unsaturated Porous Media, K-Spar, Inc., Rigby, Idaho, December, 2002.

20. Sullivan, T.M., 1996, DUST Disposal Unit Source Term, Data Input Guide, NUREG/CR-6041, BNL-NUREG-52375, Brookhaven National Laboratory, Upton, New York.

21. 1994 Final Supplemental Environmental Impact Statement, Closure of the Dawn Mining Company Uranium Millsite.

22. Final Environmental Impact Statement for the Commercial Low-Level Radioactive Waste Disposal Site, August 2004. http://www.doh.wa.gov/ehp/rp/waste/final-eis.htm

2.5.5 State of Colorado Experience with Waste Repository Covers and Caps

Lawrence J. Bruskin, P.E.
Steve Tarlton, P.E.

Colorado Department of Public Health & Environment
Hazardous Materials and Waste Management Division
Denver, CO

Executive Summary

The Colorado Department of Public Health & Environment (CDPHE), Hazardous Materials and Waste Management Division (HMWMD) is tasked with regulating waste containment facilities for three regulatory programs; solid waste, hazardous waste, and radioactive waste. Similar to other states, waste repositories in Colorado have historically used engineered barrier cover systems to prevent moisture percolation into the waste below. However, based on research as well as actual experience in Colorado over the last decade, the HMWMD now encourages the use of water balance covers and actively discourages the use of near surface engineered barriers for all programs, where applicable. These water balance covers generally take the form of a prepared subgrade above the waste, a capillary break and/or biota barrier above the subgrade, a monolithic loose soil layer capable of storing the required moisture and supporting vegetation above the capillary break, and a stand of native vegetation. As an example of water balance cover performance in Colorado, percolation is measured in three pan lysimeters constructed directly beneath the Shell Disposal Trenches RCRA-Equivalent Cover (Shell Cover) at the Rocky Mountain Arsenal (RMA). According to the RMA Remediation Venture Office (2010a)[1], for the 12-month period ending March 2010, percolation measured in the three lysimeters averaged 0.0018 mm, which is well below the 3 mm to 4 mm percolation that can be expected for composite barrier covers, as presented by Benson (2010)[2].

Climate

Within most of the state, Colorado is ideally suited for water balance covers because of its climate. Geographically, Colorado can be divided into three approximate N-S trending zones across the state. From east to west, the three zones are colloquially called the "eastern plains", "central mountains", and the "western slope". The climate of the eastern plains and western slope are somewhat similar, while the central mountains, with peaks ranging above 14,000-ft elevation, are unique. Fortunately, nearly all of Colorado's hazardous and radioactive waste repositories are located on the eastern plains or western slope.

The eastern plains and western slope have a semiarid climate, characterized by low humidity, moderate precipitation (P), and significant potential evapotranspiration (PET). For example Denver, which is located along the western extent of the eastern plains, as well as Grand Junction, which is located on the western slope, have an average annual P of about 400 mm (16-in) and 230 mm (9-in) respectively, with each area having an average annual PET of about 1,500 mm (60-in), thus an approximate P/PET no greater than about 0.25. The climatic conditions in the eastern plains and western slope of Colorado are therefore very favorable for water balance cover success. From a national perspective, it appears that many other areas of the U.S., including the western high plains and intermountain west, would also have favorable climatic conditions for successful water balance covers.

Past Experience

During the 1990's, seven UMTRA Title I cells with low permeability radon barriers were designed and constructed in Colorado. These are: Durango, Grand Junction, Gunnison, Maybell, Naturita, Rifle, and Slick Rock. In addition, three Title II sites, Durita, Maybell, and Uravan, were stabilized using cover designs based on the approved Title I sites. The Durango site was the first Title I cell to be completed in 1991, while the Grand Junction disposal site was completed in 1998. Therefore, the age range of the various disposal cells (approximately 12 to 20 years) allows some interpretation of the current conditions as well as an extrapolation of long term conditions that can be expected.

With the exception of the Durango cell, which was finished with a rock-soil matrix top, the other six Title I and three Title II disposal cells in Colorado were completed with large riprap on the top deck designed specifically for erosion protection. The U.S. Department of Energy (DOE), the agency responsible for operation and maintenance activities at all of the completed UMTRA Title I disposal cells, inspects and reports annually on each cell's condition. According to the most recent inspection and monitoring report by DOE (2010)[3], four of the seven Colorado Title I cells (Durango, Grand Junction, Gunnison, and Maybell) currently require some active maintenance with respect to vegetative growth.

The maintenance typically consists of removal of undesirable vegetative species, noxious weeds, or deep-rooted vegetation. Maintenance is required because vegetation is capable of transpiring water out of soil, including low permeability radon barriers, thus affecting its ability to limit radon emanation and infiltration. Numerous researchers, including Waugh and Richardson (1997)[4], Smith (1999)[5], Waugh (2004)[6], and Breshears and others (2005)[7] describe the negative effects on low permeability barrier layers due to root penetration or macropores left by decomposing plant roots. At the Durango and Gunnison cells, the HMWMD has observed several different types of woody plants (e.g., trees and shrubs) that have grown on the finished cover surface or side slope between the riprap. The DOE has removed the trees as part of ongoing maintenance activities, but unwanted vegetation on disposal cell covers is clearly a long-term concern for the HMWMD.

Full Scale Water Balance Cover

The Rocky Mountain Arsenal (RMA), located in Commerce City, Colorado, about 11 miles (18 km) northeast of downtown Denver, was built during World War II and operated until 1982. During this time, the facility manufactured chemical warfare agents and other incendiary munitions for military use. During the 1950s and 1960s, part of the facility was leased to private companies, such as Shell Oil Company, for manufacturing herbicides, pesticides and other agricultural chemicals. Waste disposal practices during this time resulted in contaminated soil, sediment, surface water, groundwater, and structures. The remedy selected for cleanup in the 1996 RMA Record of Decision (ROD) required large contaminated unlined land areas to be covered with a "RCRA-equivalent cap". This terminology implied that the cover to be constructed would have to be consistent with Subtitle C (hazardous waste) landfill requirements, or the equivalent protection with respect to percolation must be attained.

The U.S. Army along with Shell Oil Company and the U.S. Fish & Wildlife Service (collectively called the Remediation Venture Office [RVO]) elected to implement a water balance cover instead of a barrier cover. In order to demonstrate the "equivalency" to a Subtitle C cover as required in the ROD, soil cover test plots were designed and constructed. The four test plots were constructed using different soil moisture storage layer thicknesses (42, 48, and 60-inches) and two different soil textures (based on percent passing the No. 200 sieve). Vegetation, consisting of native grasses and forbs designed to emulate the surrounding natural short-grass prairie, was established over a three year period followed by a one year test period. Percolation was measured in each plot during the test year using a 30-ft x 50-ft pan lysimeter. The results of the test year indicated that all four cover configurations had acceptable percolation performance in comparison to the regulatory percolation criterion of 1.3 mm per year, so any of the test plot configurations would be acceptable for full-scale application. The RVO ultimately selected for implementation a 48-in thick cover with a capillary break.

The 17-acre Shell Cover was the first RCRA-Equivalent cover constructed at RMA, and was completed in September 2007. The Shell Cover design was slightly different than the remaining RMA RCRA-Equivalent covers that have since been constructed, as the Shell Cover contains a 6 oz/sy geotextile fabric between the chokestone/biota barrier and the cover soil. The Shell Cover also contains three pan lys meters within the cover footprint for compliance monitoring. Nests of time domain reflectometer (TDR) moisture probes along with temperature probes were also installed in the Shell Cover near each lysimeter. By agreement, moisture monitoring devices were only installed in the Shell Cover. Since the Shell Cover was completed, more than 425-acres of additional RCRA-Equivalent covers were constructed at RMA (without the geotextile), along with 17 additional pan lysimeters for measuring compliance. In lieu of the geotextile, a nominal 2-in layer of washed squeegee (relatively uniform, clean, fine gravel) was placed between the chokestone/biota barrier and cover soil to form the required capillary barrier.

At RMA, information from the moisture probes in the Shell Cover will not be used for compliance evaluation, but rather to assist in the selection of an appropriate maintenance activity or corrective action should the annual percolation measured approach or exceed the regulatory percolation criterion of 1.3 mm per year. It is the HMWMD's position that large-scale lysimeters are the most representative and accurate method for measuring cover percolation against an enforceable regulatory criterion. Conversely, the HMWMD does not support the use of moisture probes for compliance assessment because of the high degree of uncertainty and imprecision associated with percolation estimates based on water content data obtained from the probes.

To date, the percolation values measured within the lysimeters on the Shell Cover are extremely low. The total percolation for the 12-month period ending March 2010 showed the three lysimeters had percolation of 0.0 mm, 0.0001 mm, and 0.0053 mm, for an average value of about 0.0018 mm per year, well below the regulatory criterion of 1.3 mm per year. This is even more impressive considering that the Shell Cover has been completed and vegetation established for about three years. By agreement with the RVO, enforcement of the compliance standard does not begin until the sixth year after construction completion. The RVO as well as HMWMD are optimistic that the measured percolation in the future will be no greater or even smaller than the most recent values once vegetation becomes fully established in the next several years.

All of the RCRA-Equivalent covers at RMA require inspections and maintenance at regular intervals as well as non-routine maintenance when necessary. According to the most recent Shell Cover report by the RVO (2010b)[8], the only problems observed included some minor soil cracking attributed to consolidation of the cover soil following irrigation, tire tracks created by tractors that mow the cover, relatively small areas of poor vegetation around two of the lysimeters, and a few isolated burrowing animal holes. Deep-rooted woody plants or trees similar to those found at the Colorado UMTRA cell covers have never been problematic at RMA.

Conclusions

There is no evidence of cover failure for any of the seven Title I or three Title II cells constructed in Colorado with rock-armored tops. Items such as elevated radon flux, seeps, excessive erosion, or slope instability have never been observed or attributed to the performance of final covers in Colorado. However, multiple lines of research and investigation have shown that unwanted vegetation will quickly establish between the large rocks on the cover surfaces, requiring ongoing vigilant maintenance. This has already happened at the majority of the Colorado disposal cells. The preferred alternative to rock covers in Colorado are water balance covers, with beneficial vegetation designed to mimic the natural surrounding environment. Based on the percolation data collected to date, water balance covers in Colorado appear to offer at least an equivalent level of cover performance compared to engineered barrier designs. In fact, the HMWMD postulates that water balance covers can provide a superior level of long-term performance in Colorado. Even if percolation performance were equal, water balance covers provide an excellent way to restore the natural ecosystem, are usually easier to construct compared to engineered barrier covers, and usually cost less to construct and maintain when compared to barrier covers.

References

[1] Rocky Mountain Arsenal Remediation Venture Office, 2010a, *Shell Disposal Trenches RCRA-Equivalent Cover Soil Cover Moisture Monitoring System Data Evaluation Summary January 2010-March 2010, Rev. 0*, prepared by Tetra Tech EC, Inc., June 23, 2010

[2] Benson, C.H., 2010, *Introduction to Water Balance Covers*, presented at the Workshop for Water Balance Covers From El Paso to Port Arthur; Design, Modeling, Construction, and Monitoring, March 30-April 1, 2010, Austin, TX

[3] U.S. Department of Energy, Office of Legacy Management, 2010, *2009 Annual Site Inspection and Monitoring Report for Uranium Mill Tailings Radiation Control Act, Title I Disposal Sites*, LMS/S05884 http://www.lm.doe.gov/pro_doc/guidance_reports.htm

[4] Waugh, W.J and Richardson, G.N., 1997, *Ecology, Design, and Long-Term Performance of Surface Barriers: Applications at a Uranium Mill Tailings Site*, in Barrier Technologies for Environmental Management: Summary of a Workshop, National Academy Press, Washington, D.C. http://www.nap.edu/openbook.php?record_id=5539&page=54

[5] Smith, G.M., 1999, *Evolution of Disposal Cell Cover Design Used For Uranium Mill Tailings Long-Term Containment*, U.S. Department of Energy, Office of Legacy Management

[6] Waugh, W.J., 2004, *Design, Performance, and Sustainability of Engineered Covers for Uranium Mill Tailings*, Proceedings of the Workshop on Long-Term Performance Monitoring of Metals and Radionuclides in the Subsurface: Strategies, Tools, and Case Studies, U.S. Geological Survey, April 21-22, 2004, Reston, VA http://www.cistems.fsu.edu/PDF/waugh.pdf

[7] Breshears, D.D., Nyhan, J.W., and Davenport, D.W., 2005, *Ecohydrology, Monitoring and Excavation of Semiarid Landfill Covers a Decade after Installation*, Vadose Zone Journal, Volume 4, August 2005, pp 798-810 http://ag.arizona.edu/research/breshears/publications.htm

[8] Rocky Mountain Arsenal Remediation Venture Office, 2010b, *Annual Covers Report 2009, Rev. 0*, prepared by Tetra Tech EC, Inc., March 9, 2010

2.5.6   The Navajo Nation Environmental Protection Agency Experience with Former Uranium Mill Sites

Stephen A. Austin, Senior Hydrologist

Navajo EPA Water Quality/NPDES Program
nnepawq@frontiernet.net

The Navajo Nation Environmental Protection Agency (Navajo EPA) has been charged with the task of protecting public health and the environment on the Navajo Nation. Oftentimes, this task includes working closely with federal agencies to ensure that threats to these values are properly addressed. Such is the case with the five former uranium ore processing mills located on the Navajo Nation.

During the Cold War, uranium ore was mined throughout the Navajo Reservation. Uranium was also processed on the reservation at Monument Valley and Tuba City, AZ; Mexican Hat, UT; and Church Rock and Shiprock, NM. Four of these five former mills are now Uranium Mill Tailings Radiation Control Act (UMTRCA) Title I sites. Church Rock is a Title II site. Disposal cells were constructed during the 1980s and 1990s at the Shiprock, Tuba City, Church Rock, and Mexican Hat sites. These cells contain the tailings, buildings, and other contaminated materials from the mills and offsite vicinity properties left behind when production ceased. Contaminated material from Monument Valley was disposed of at the Mexican Hat facility.

Of the five former uranium mill sites, Shiprock has received the most attention due to its location and complex hydrogeology. The 76 acre disposal cell is located within the Navajo community of Shiprock and is situated on a 50-60 foot high terrace immediately south of the San Juan River floodplain. The cell is on a 30-40 foot thick layer of unconsolidated alluvial gravels and cobbles underlain by Mancos Shale bedrock. Ground water contamination from the disposal cell was not a primary concern when it was constructed in 1986. Thus, it was not lined and the cover was designed more to keep radon from escaping than to keep water from infiltrating into the cell. It is composed of three layers: a 72 inch compacted silt/clay radon barrier; a 6 inch sand and gravel capillary break; and a 12 inch rock erosion control layer. Slopes on the cell vary from 2-4 percent on top to 20 percent on the sides.

Ground water in both the terrace and the alluvial aquifer was contaminated with high levels of ammonia, manganese, nitrate, selenium, strontium, sulfate, and uranium during the milling operations in the 1950s and 1960s. Most of this contamination was from the disposal of raffinate in several acres of unlined ponds located on the terrace. By the early 1960s, contaminated ground water had surfaced in two nearby drainages and in seeps along the terrace up to one mile to the northwest. All of these sites were still flowing into the early 2000s and some are still a concern today.

DOE oversees the ground water remediation at this site under the authority of UMTRCA. They are required to meet the remediation standards promulgated by US EPA in 40 CFR 192. Navajo EPA has worked with Navajo UMTRA and Navajo DOJ to ensure that the ground water remediation strategies implemented by DOE are protective of human health and the environment on the Navajo Nation. The primary role that Navajo EPA has played in this process has been to review and comment on the numerous documents and data produced by DOE and their consultants.

During the review process, several disagreements and concerns arose regarding site characterization and remediation goals and strategies. These included:

1) Is the terrace ground water source natural or anthropogenic? If it is primarily anthropogenic, is it all mill-related or did another source such as irrigation significantly contribute to the volume?

2) Is the contaminant source primarily mill-related or did leaching of naturally occurring contaminants found in the Mancos Shale significantly contribute to the plume?

3) How often, when, and which ground and surface water monitoring sites should be monitored?

4) What is the best strategy for reducing exposure risk where the ground water has surfaced in nearby drainages and seeps?

5) Have the conditions set forth in 40 CFR 192 for the use of alternative ground water remediation standards been met? Do these standards have to be specifically defined and, if so, what are they?

6) Will the remediation strategy of pumping dry the eastern part of the terrace work? Will the contaminated material left in the formation be remobilized if natural recharge does occur or another anthropogenic water source is introduced in the future?

7) How will natural flushing of the western part of the terrace take place if natural recharge does not occur?

8) What is the drainage of contaminated water from the disposal cell and how much will it contribute to the ground water of the terrace and floodplain? Will it be a continuing source of contaminated water given that studies have shown atmospheric water has infiltrated through the cover and saturated at least part of the underlying tailings? Can the remediation goals be met if this is a continuing source of contamination?

9) Given the close proximity to the community of Shiprock and the San Juan River and the fact that it may be a continuing source of contamination to the ground water, can the disposal cell be removed as is proposed for the Moab site?

10) Does DOE need to obtain a NPDES permit under the Clean Water Act for the discharge of any contaminants not regulated under the Atomic Energy Act?

Though the relationship between Navajo EPA and DOE has been somewhat antagonistic in the past, the current oversight by the Office of Legacy Management has lead to more cooperation and acceptance of our concerns. Greater effort has been made to listen to our concerns and to include us in the decision making process. Much work still needs to be done in order to adequately address all of the issues listed above. Our improved relationship with DOE gives us hope that we can work together more closely in the future to meet our goal of protecting human health and environment on the Navajo Nation.

2.5.7 New York State Department of Environmental Conservation's Experiences with Solid and Hazardous Waste and Low Level Radioactive Waste Containment Systems – A Lessons Learned Perspective

Robert Phaneuf, Tim Rice, Dave O'Hehir, John Mitchell, and Jaime Lang

Division of Solid & Hazardous Materials
New York State Department of Environmental Conservation
Albany, New York

This paper provides a summary of New York State's experiences with assessing engineered barriers for environmental containment systems from a lessons learned perspective. The reader will be given an overview of State's solid and hazardous waste landfill barrier/liner system requirements. Information will be summarized from facility monitoring data collected over the last 2 decades that illustrates that our engineered liner systems are in fact highly effective. However, the data also substantiates the need for ongoing containment system maintenance and monitoring. The reader will be apprised of the importance and benefits of proactive internal containment system monitoring of witness zones designed into the containment system and how this approach can help ensure adequate containment accountability. The success of this approach has been verified by actual groundwater quality monitoring analysis at the state's double-lined landfills – establishing no groundwater impacts attributable to liner system leakage over a period in excess of 22 years. A brief overview of State's Low-Level Radioactive Waste (LLRW) Disposal Facilities: Certification of Proposed Sites and Disposal Methods Regulations (6 NYCRR Part 382) will be given and a number of case histories at LLRW disposal facilities that existed in the State prior to the development of these regulations, or that were developed by a federal agency not subject to the State's current regulations that lend support to some of the State's current LLRW regulatory requirements. In conclusion the paper's authors will attempt to combine these lessons learned from both the solid waste and LLRW program areas in a manner that will be helpful to Workshop participants.

Introduction to a State's Lesson Learned from a Barrier Requirement Perspective – Evolution of the State's Solid Waste Landfill's Liner Requirements

Most of New York State's experience with monitoring and performance assessments of modern landfill liner systems has been associated with the State's solid waste landfills that were required to be double-lined and closely monitored on a routine basis since before 1988.

In the late 1980s the Department experienced difficulties with permitting of new and needed solid waste landfills. The public's mistrust and lack of confidence in the modern waste disposal facilities at that time was largely over the concerns for adequate groundwater protection. What was determined as being needed for these disposal facilities were that conservative settings should be sought where groundwater is constrained to a large degree by the natural geology in combination with rigorous liner system requirements whereby we could be assured that groundwater resources would be protected by effectively isolating the waste from the environment. Out of these concerns came the promulgation of the December 31, 1988 version of state's solid waste management regulations (6 NYCRR Part 360) that required that all solid waste landfills be lined with double composite liners and dual leachate collection and removal systems. With the past regulatory history and failed attempts to ensure adequate groundwater protection associated with improper disposal facility design and operation, these new regulatory requirements went well beyond the "belts and suspenders" of just the double-liner system. A comprehensive regulatory approach was put into place for governing the siting, design, construction, operation, closure and post-closure of all solid waste disposal facilities.

Since 1988 the State's landfill regulations have contained prescriptive requirements for double composite liners and dual leachate collection and removal systems and composite final cover system designs. Over the years of implementing these regulations the important connection between construction quality and acceptable containment performance has been realized. Today the regulations have construction quality control and quality assurance requirements, along with final construction certification requirements, that verify acceptable containment system performance prior to the facility being authorized for waste acceptance. On top of all this the State's regulations also contain specific operational requirements incorporating on-going comprehensive environmental monitoring programs including landfill liner system performance and routine containment system assessment and on-going maintenance requirements to ensure continued acceptable performance (see 6 NYCRR Part 360-2 Landfills at this web link for the current version of the State's landfill regulations: http://www.dec.ny.gov/regs/2491.html).

This on-going monitoring and maintenance is not only required during operations but also is required to continue on into the post-closure period. The regulations require a comprehensive and proactive approach to ensuring compliance with acceptable upper liner performance thresholds that are used as a barometer of sorts to predict internal containment system performance problems/issues prior to them posing significant threat to outward releases of contaminants. This proactive internal containment system monitoring approach has contributed to these modern disposal facilities not having any detrimental impacts to groundwater quality that are directly attributable to leakage from their engineered containment systems. This conclusion is supported by over two decades of groundwater quality monitoring at these facilities. This type of containment system monitoring requires that facility operators react in a timely manner to problems as they present themselves. In order to implement corrective operational maintenance procedures it is important to ensure that these containment systems be designed with the ability to access and maintain key drainage system components within the landfills double liner system.

With respect to groundwater monitoring it s important to understand that due to the siting requirements for these landfills requiring good containing geologic formations, many of the State's landfills are sited in areas of relatively shallow groundwater tables. The regulations a low for landfill liner systems to be placed below the groundwater elevation providing that the landfill is constructed with a pore pressure relief system (a lower drainage layer beneath the double lined landfill designed to minimize hydrostatic uplift pressures or the liner system). These basal drainage systems have been installed at about 70 percent of the State's landfills and are used as a groundwater monitoring point as wel

The Department posts the annual reporting data that is collected pursuant to the State's solid waste regulations on the Department's ftp site. The direct link to this annual report data is: ftp://ftp.dec.state.ny.us/dshm/SWMF/.

By reviewing of the State's solid waste landfill liner performance data one can see that a high degree of containment efficiency for the upper liner system has been exhibited (see Figure 1 below). When one applies this upper liner system efficiency to the lower liner with the actual flows being monitored in our leak detection zones it is easy to see why these landfills pose little or no threat to groundwater quality.

The evolution of the Department's liner system requirements is still on-going and the proposed changes to these existing liner requirements are being drafted that will:

- Enhance long-term liner system performance by building on the improved performance we have experienced from our current monitoring and maintenance programs;
- Greatly reduce construction related defects in the double liner system, helping to sustain better long-term containment system performance through improvement to the existing construction quality assurance requirements; and,
- Contain the needed flexibility to allow deviations from prescribed construction materials such that new and improved materials and test methods that are constantly evolving can be used that add quality and enhanced long-term performance.

The State's hazardous waste regulations governing hazardous waste landfills, 6 NYCRR Part 373, have similar requirements to that of the solid waste landfills however, these regulations for the most part mimic the federal hazardous waste regulations for land disposal facilities and tend not be as specific as those set forth in 6 NYCRR Part 360 as they relate to containment system design, construction and operation. These regulations can be accessed from this web link: http://www.dec.ny.gov/regs/2491.html.

New York State Regulation of Low-Level Radioactive Waste (LLRW) Disposal Facilities: Certification of Proposed Sites and Disposal Methods (6 NYCRR Part 382)

The State's LLRW land disposal facility requirements stipulate that LLRW disposal facilities - must be sited, designed, operated, closed, and controlled after closure to reasonably assure that exposures to humans are within the limits established in the performance objectives. Today's regulations require that concentrations of radioactive material which may be released to the general environment must not result in an annual dose exceeding an equivalent of 25 millirems to the whole body, along with a number of other performance objectives (see sections 382.10 through 382.15).

The regulations stipulate that releases of radioactivity in effluents to the general environment must be maintained as low as reasonably achievable.

The siting requirements for LLRW disposal facilities in the State are such that they prohibit the deposition of LLRW in shallow below grade repositories. This is largely due to the fact that the siting requirements seek out geologic formations in the State where groundwater flow is restricted within good containing formations where groundwater flow is extremely slow. These geologic formations due to their characteristics predominately also exhibit shallow groundwater elevations where siting a repository in such a geologic unit raises issues with groundwater seepage. As such the regulations encourage above grade repositories such that barriers can be designed and monitored to ensure that the above stated performance standards can be met. These requirements also allow for the disposed LLRW to be accessible in the event that it is determined necessary or advantageous to recover these materials. The fact that the state experienced two heavily publicized failures of early radioactive waste shallow-land disposal sites was also taken into account when drafting these regulations, see the case histories discussed below.

The other disposal options available under the State's LLRW regulations are deep underground mining applications. Here the disposal site must not be located in a geologic unit where subsurface hydrogeologic processes, such as dissolution can occur with such frequency and extent so as to adversely affect the ability of the underground mined repository to meet the performance objectives mentioned above. The regulations emphasize that geologic formations and hydrogeologic features of the site must be determined to be safe for construction, operation, and closure of an underground mined repository. The geologic unit where the underground mined repository is to be located must have adequate thickness and lateral extent to contain the disposal units and must not contain features, such as faults, fractures, or other discontinuities, which could adversely affect the underground mined repository's ability to meet the regulatory performance objectives of Sections 382.10 - 382.15. Obviously, these limitations make selection of this option less likely than the above grade alternative.

Other siting considerations contained in the State's LLRW regulations require that the site must be located in an area where subsidence and related impacts or where any instability of the geologic unit will not adversely affect the underground mined repository's ability to meet the regulatory performance objectives.

In order to best ensure that the regulatory performance objectives of Sections 382.10 - 382.15 are capable of being met for the long-term the regulations require that the land on which the disposal area is located is to be owned by the State, leased to an operator and will ultimately be placed in the custodial care of the State for the institutional control period. Provisions in the regulations also require the establishment of trusts and escrow funds in order to provide for necessary financial assurance that the operator and State will have adequate monies to ensure that long-term care responsibilities during the lifespan of the facility can be conducted. As we learn more about that service life and maintenance needs of our long-term LLRW repositories it strengthens the need for performing full cost accounting for long-term post-closure needs to best inform disposal decisions made today ensuring that we do not unduly burden future stakeholders financially.

Radioactive Site Case Histories

Radioactive Site Case Histories

West Valley State-licensed Disposal Area (SDA)
http://www.nyserda.org/programs/West_Valley

The West Valley State-licensed Disposal Area (SDA) was one of the first commercial radioactive material disposal sites, along with Maxey Flats (KY), Sheffield (IL), Beatty (NV), Barnwell (SC), and Richland (WA). It is located on the larger 3,345 acre Western New York Nuclear Service Center, home to the country's only operational commercial spent nuclear fuel reprocessing facility, and an NRC-licensed disposal facility that supported the reprocessing operation (see discussion below). All three were operated by Nuclear Fuel Services (NFS) under a lease from the New York State Energy Research and Development Authority (NYSERDA).

The SDA is a 15 acre State regulated pre-Part 61 shallow land disposal facility which received radioactive wastes from 1963 through 1975. There are two areas within the SDA, north and south, each consisting of seven disposal trenches. It was licensed to receive by-product material, source material, and special nuclear material. Therefore it contains considerable waste not compliant with Part 61 era LLRW regulations.

Geologically, it is located in a very tight glacial till clay layer with a thin surface layer (a few feet) of weathered clay containing intermittent cracking. The original final cover system consisted of a contoured grass covered compacted native clay barrier layer placed over the trenches to limit infiltration and isolate the wastes. This simple earthen cover proved to be largely ineffective to preventing infiltration. The clay barrier was subjected to the impacts from frost action and periodic desiccation and cracking, and the waste in the trenches settled causing the barrier layer to settle. Subjecting the barrier to all of these conditions increased the barrier permeability and allowed precipitation to more easily infiltrate the trenches.

The tight geologic formation in which the unlined trenches are constructed turned out to have both advantages and disadvantages. As originally envisioned, the geologic formation effectively contained the waste, but it also contained any water that seeped into the burial trenches. This resulted in increased hydraulic heads building or mounding in the waste trenches overtime. In 1975, when accumulated liquids within the northern set of trenches reached the more permeable overlying weathered clay layer, the contaminated liquids migrated through it, day-lighting along adjacent slopes, the classic "bathtubbing effect". In an attempt to contain the contaminated seepage several episodes of pumping leachate out of the trenches occurred over the next few years. This leachate was treated and released through the site's liquid radioactive waste treatment system.

To try and stop this infiltration in the southern set of trenches efforts included excavating and removing a sand lens along with installation of a concrete cutoff wall for trench 14 – these efforts were only partially successful. Given continued leachate level increases, in 1992, an upgradient slurry wall was installed along the upgradient side of trench 14 in an effort to divert groundwater flows around the SDA. As part of that effort an exposed VLDPE geomembrane cover was placed on top of existing soil cover system for trenches 14, 13, and half of trench 12 as an interim remedy under a RCRA Consent Order. This combination of slurry wall and geomembrane cover proved to be successful in stabilizing the leachate levels in these trenches and the cover was repeated over the rest of the 15 acre site using a geosynthetic composite geomembrane (XR-5). To control the storm water runoff from the exposed geomembrane detention basins with low-flow and overflow outfalls were designed as integral components of the exposed interim geomembrane cover system.

NYSERDA is currently developing studies for the SDA to investigate the feasibility of partial waste exhumation, and to further refine long-term erosion modeling. They have also developed long term monitoring plans for site erosional impacts, slope stability, and environmental media.

West Valley NRC-licensed Disposal Area (NDA)
http://www.wv.doe.gov/

The NRC-licensed disposal Area (NDA) is located adjacent to the above mentioned SDA. This facility was created solely to support the NFS Spent Fuel Reprocessing operations, operating from 1966 through 1986. NFS wastes consisted primarily of the cut up and acid leached sections of fuel elements (hulls) from the reprocessing operation. It also received other related wastes including tanks with "spent" solvents used in the fuel reprocessing. In addition, NFS buried a batch of fuel rods received from a federal reactor that were damaged too badly to be able to be processed with site equipment. In its last years of operation it received waste from United States Department of Energy (DOE) site remediation efforts.

In 1983 groundwater monitoring discovered that solvent was leaking from the site that contained plutonium and other contaminants. This resulted from the same "bathtubbing" effect that plagued the SDA. In this case infiltrating groundwater and precipitation accumulated in the holes containing the solvent waste tanks, allowing the solvent to "float" to the top of the water table and move laterally through the weathered clay.

The interim solution was to locate and remove the offending solvent tanks and install a down gradient collection trench. Water accumulating in this collection trench was pumped to the sites liquid radioactive waste treatment system. This interim measure was sufficient to stop the contaminant release. Continued monitoring supports the conclusion that no recurrence of the solvent/plutonium contamination release has taken place.

On-going monitoring of the NDA determined that groundwater and precipitation seepage into the NDA was resulting in increased water levels in the NDA's trenches and may have contributed to a radiologically contaminated seep on the surface of the site. In 2009, as an interim measure under a RCRA consent order[1], the DOE successfully used the SDA interim measures approach to stabilize the water infiltration problem at the NDA. Both sites have now stabilized their water infiltration problems with application of an interim exposed geomembrane and upgradient slurry walls to divert groundwater around the disposal areas. In both the case of the SDA and NDA interim landfill covers ongoing periodic monitoring and assessment of the mechanical properties of exposed geomembrane cover will take place.

Long-term Considerations for both the SDA and NDA [2, 3]
http://www.westvalleyeis.com/finaleis.htm

The EIS process for West Valley site is unique in that it is using a "phased" approach to decommissioning. In Phase 1, with the exception of the two disposal areas and the high-level waste tank complex, all of the infrastructure and much of the site contamination is being addressed, primarily through removal. A series of ongoing studies are being developed to take place during Phase 1 to inform the Phase 2 decisions on how to address the tanks, a stabilized groundwater plume, and disposal areas. These studies will focus primarily on environmental processes (erosion, water infiltration) and the performance of engineered barriers and institutional controls. These are the issues where high levels of uncertainty (and significant differences of opinion) remain. The ultimate decision will be whether to remove these areas, close them in place, or some combination thereof.

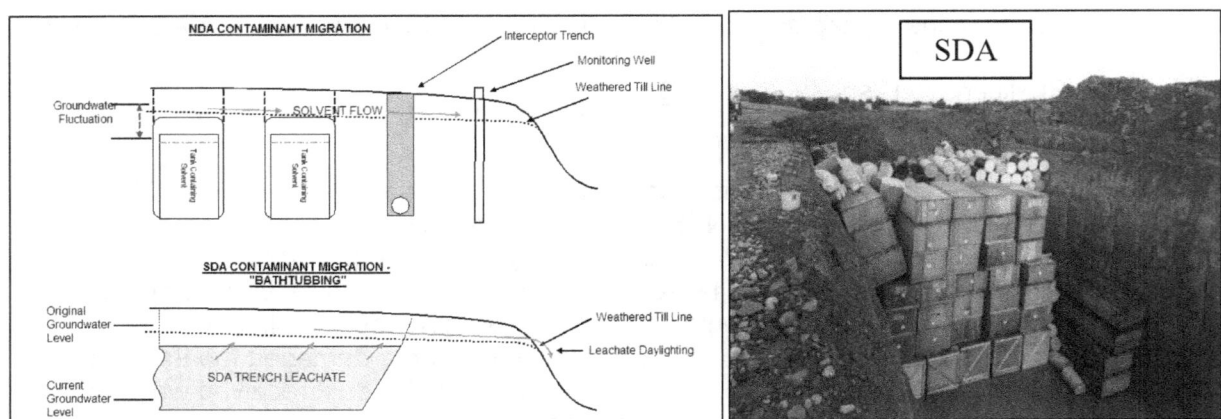

Cornell Radiation Disposal Site (RDS)
http://www.eco.cornell.edu/DSC/rds.cfm

The Cornell University Radiation Disposal Site is an approximately one acre radioactive disposal facility (shallow land burial) that was authorized pre-Part 61. The RDS is located in a seasonal wetland, is underlain by about 10 to 15 feet of impervious glacial till with high clay content and 20-25 feet of highly fractured shale bedrock, transitioning to more competent bedrock at about 40 feet. From 1956 to 1978 the facility was utilized for the disposal of radiologically contaminated animal carcasses, waste from animal care, scintillation cocktai s and short lived isotopes. Waste materials were disposed of in slit trenches 6 to 12 feet deep. In hindsight, the hydrogeological setting for the RDS was less than ideal. By siting the RDS in a wetland and excavating the trenches through the tight clay layer into the fractured shale a contaminant pathway was openec allowing surface water to infiltrate through the waste mass and carry contaminants on into the fractured shale.

In 1994 remedial investigation efforts identified a groundwater p ume extending 2500 feet downgradient of the RDS, being predominantly located in the fractured shale layer. The p ume contained strontium-90, tritium and carbon-14 at relatively low levels and volatile organic compounds, primarily paradioxane from the disposal of scintillation fluids. In 1996-97 as part of efforts undertaken pursuant to a consent order[4] an interim cover, gas venting system, and surface water drainage and erosion control measures were installed. These efforts reduced the hydraulic head within the waste trenches. but could not by themselves fully address the groundwater plume nor stop all water from moving through the waste mass. Thus, in 2001 a groundwater recovery system was installed as an additional interim measure to control the migration of the volatile organic plume. The discharge for the paradioxane treatment system contained low-concentrations of incidental radioactive material (H-3, C-14, and Sr-90) and was regulated by a radiological discharge permit.

In accordance with a 2002 Record of Decision[5] a final remedy consisting of improved surface water drainage, a more robust multi-layered cover system placed on top of the existing interim cover system, and a grout curtain and slurry wall was installed to isolate the disposal area from the environment. Subsequent to completion of the remedy in 2004 it was determined that the groundwater recovery system had met the remedial objective for paradioxane and it was shut down.

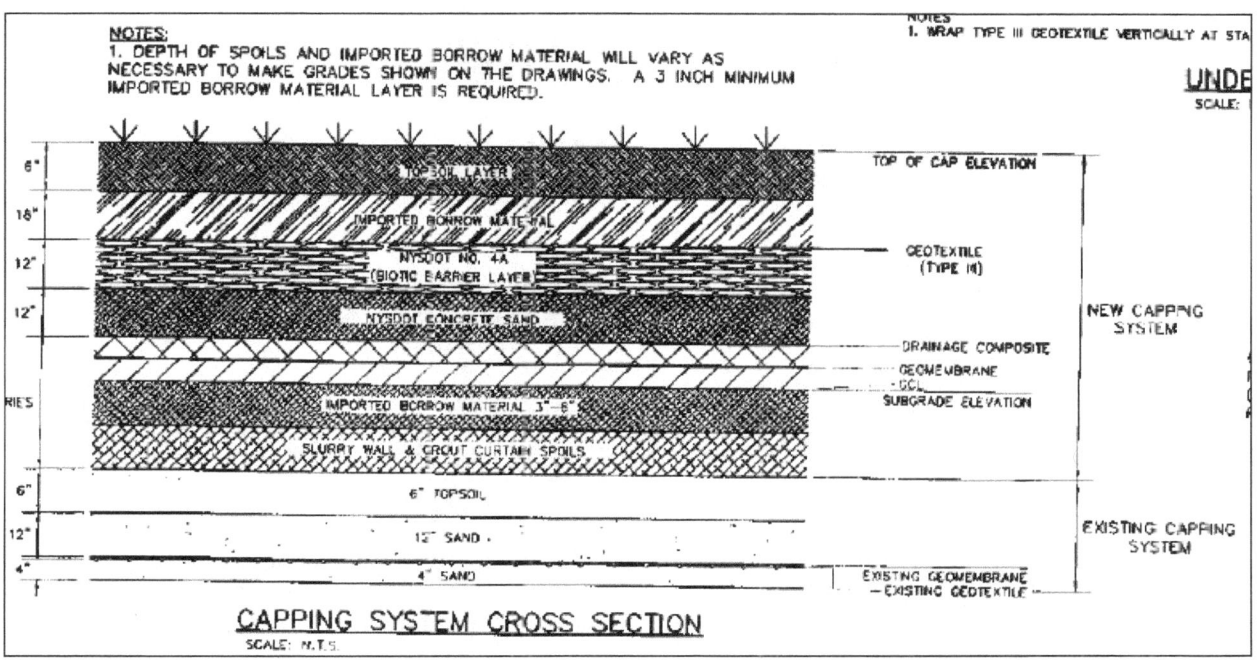

CAPPING SYSTEM CROSS SECTION
SCALE: N.T.S.

Currently the site is being managed under active monitoring and maintenance. Cornell performs routine sampling, inspection and monitoring and submits required reports to the Department. The Department performs regular inspections and reviews data to ensure proper performance of the site. Based on continued performance monitoring Cornell is looking to move forward to a long term monitoring plan for the site.

Niagara Falls Storage Site, Interim Waste Containment Structure (IWCS)
http://www.lrb.usace.army.mil/fusrap/index.htm

The federally owned 191 acre Niagara Falls Storage Site (NFSS) was originally part of a 7,500 acre World War II explosives manufacturing plant called the Lake Ontario Ordnance Works (LOOW). The IWCS occupies 10 acres of the 191-acre NFSS. Between 1943 and 1954, the Manhattan Engineer District (MED) and the Atomic Energy Commission (AEC) brought radioactive residues and wastes to the LOOW site. These residues, which make up a large portion of the world's supply of Ra-226, were mostly residues from extracting uranium from various ores for the production of the atomic bomb. Given the origins of the waste, the site is managed under the Formerly Utilized Sites Remedial Action Program (FUSRAP).

A cluster of building basements were in part used for waste confinement. The IWCS outer perimeter is formed by a dike and cutoff wall, each constructed of compacted clay and incorporated into the finished structure. The cutoff wall extends a minimum of 18 in. (45 cm) into an underlying gray clay unit. The gray clay unit and the cutoff wall/dike serve as adsorption barriers to vertical and horizontal migration of contaminants from the structure.

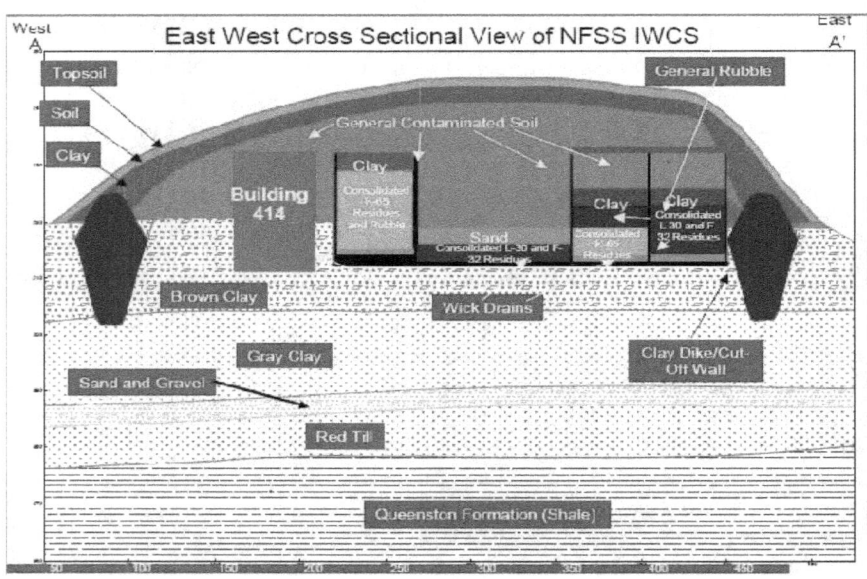

An engineered, compacted clay cover is placed immediately over the wastes and extends beyond the perimeter dike, completely enclosing the containment structure. This clay cover is the principal barrier against moisture intrusion and radon emanation. The clay layer is covered with a surface layer of loosely compacted, 18-in.-thick soil cover and topsoil. This surface layer forms a protective blanket intended to maintain moisture and prevent drying that could result in the formation of tension cracks within the clay layer and to reduce the effects of frost action. It also provides a base for shallow-rooted grass. The clay layer, soil cover, and topsoil comprise the short-term closure system for the IWCS.

Environmental Surveillance activities began on the Niagara Falls Storage Site (NFSS) in 1981 under the direction of the DOE. Over the years, the Environmental Surveillance Program (ESP) for the site has evolved to demonstrate the continued containment of wastes and residues buried within the on-site IWCS and to ensure that on-site contamination does not pose a threat to human health and the environment. In October 1997, Congress transferred the responsibility for FUSRAP from the USDOE to the United States Army Corp of Engineers (USACE). In addition to investigating and remediating site contaminants at the NFSS, the USACE is responsible for maintaining the site and conducting the environmental Surveillance Program.

Media monitored at the NFSS includes: surface water, groundwater, and streambed sediment for radiological and chemical parameters, and air for radon and external gamma radiation dose.

✓ Annually 180 radon flux canisters are placed on the IWCS protective cap to measure the release of Radon-222.

✓ Radon and external gamma radiation monitors are located around the IWCS and the perimeter of the site. These detectors are exchanged twice per year.

✓ Surface water and sediment samples are taken annually and tested for radium, thorium, and uranium.

✓ Groundwater wells are sampled annually and tested for the radioactive constituents listed above as well as metals and other water quality parameters. Water level measurements are recorded quarterly from 91 wells to determine the groundwater flow directions in the upper and lower water-bearing zones.

Monitoring of all media to date has shown no release of radiological constituents.

Despite the favorable monitoring data to date it is the Department's position that the EPA's 40 CFR Part 192 uranium mill tailings standard used in the design of the IWCS was never intended to regulate such high activity radioactive wastes (the K-65 residues), and the more applicable standard would be 40 CFR 191 requiring deep mine repository disposal.

Lessons Learned: With Respect to Effective Liner Systems from Both the State's Solid and Hazardous Waste and LLRW Programs

From our experiences we have learned the following relative to effective barrier system performance:

That Routine Environmental Monitoring Allows for Effective Compliance Assessments: Routine monitoring of a disposal site and its hydrogeologic setting can effectively signal need for remedial actions and is capable of being used to evaluate the performance of the required remedial activities.

The Important Role of Geosynthetics: We learned that we cannot build an effectively compliant modern containment system for shallow land disposal facilities in New York State's humid climate, and shallow groundwater setting without the use of properly designed and specified geosynthetic materials as integral components of the environmental containment system. The use of a geomembrane installed in direct contact with a low permeability soil layer greatly improves the hydraulic containment characteristics of the barrier system, whether it is in a disposal facility liner or final cover system. Use of geotextiles as separation barriers help to ensure drainage layers do not become clogged. Use of geocomposite drainage layers afford efficient side slope drainage of the secondary drainage layer between the upper and lower liner systems on landfill liner side slopes – where use of soil drainage layers in these areas can present veneer stability issues. Geosynthetic clay liners in lieu of compacted clay barriers used in composite landfill final cover systems outperform composite covers made with compacted clay barriers over the long-term.

Construction Quality Is Directly Related to Operational Performance: Since the application of double liner systems and implementation of liner system performance monitoring – we have learned that shoddy construction is evident in early stages of operation. In numerous cases shoddy construction can go undetected even with third party construction oversight. From the literature we know that 97 percent of all geomembrane liner defects occur during liner system construction, 73 percent of which occur from drainage blanket material placement alone[6]. Thus to ensure quality construction and the best long-term hydraulic containment performance post construction electrical resistivity leak location surveys should be conducted and incorporated into the projects construction quality assurance plan from the start when geomembrane barriers are used.

Barrier Layers are Only Effective if Coupled with An Effective Drainage Layer: The function of the leachate collection and removal system is equally as important as the liner itself in ensuring adequate long-term hydraulic containment system performance. From daily monitoring of the landfills upper and lower leachate collection and removal systems we have learned that increased hydraulic head above the upper liner results in increased leachate flows of the secondary leachate collection and removal system. We have learned that the more permeable and porous the drainage layer is the lower the incidence and threat to system clogging, and that hydraulic heads on the liner system can be kept to a minimum when the system is designed properly.

Maintenance is Essential to Ensuring Long-Term Performance: We have learned that these designs need to incorporate the ability to maintain and access key portions of the leachate collection and removal system to prevent system clogging. Our monitoring data substantiates that hydraulic containment system performance is responsive to routine maintenance of the leachate collection system.

Solving One Problem May Create Unintended Consequences: We have learned that using more porous drainage materials in the drainage layer results in the need to use appropriately designed geocushion materials to protect the geomembrane liner.

Proactive Monitoring of Liner System Performance Ensures Compliance: We have learned that double-lined landfills afford an ability to proactively monitor the hydraulic functioning of the landfill's internal containment system such that remedial options can be advanced in a timely fashion, avoiding the potential of impacting the environment or groundwater resources.

Siting Decisions Sometimes Affect Long-Term Operational Needs: Use of groundwater suppression or pore pressure relief systems constructed under the double lined landfills in areas of high groundwater tables allows for safe construction of the landfill liner system below ground water elevations in certain instances. These systems need to be maintained and operated until uplift forces are demonstrated to no longer present a threat to liner system stability. If these pore pressure relief systems are shut off they may result in the need to perpetually collect liquid from secondary leachate collection and removal system to minimize leachate production in the long-term.

New and Improved Containment Materials and Test Methods are Constantly Being Developed: We have learned that new and improved construction materials and specifications are constantly being developed along with new and improved test methods for construction quality assurance and control that need to reviewed and updated to ensure the best available practices are being followed.

ALL Construction Materials Have A Service Life: We have learned that all construction materials do have their respective service lives and that the designer needs to take these anticipated service lives into consideration when designing environmental containment systems. Understanding of the construction materials degradation process can afford better long-term containment system designs for the future. For example a geomembrane in landfill liner system will outlast a geomembrane used n a final cover system due to different exposures and stresses in both applications having different affects on material degradation. The Geosynthetic Research Institute (GSI) suggests that the typical service life of a HDPE geomembrane liner beneath a solid waste landfill could be on the order of 700 years. Whereas the service life of that same geomembrane used in a final cover system may be on the order of only 170-200 years. GSI suggests that service life determinations of geosynthetic materials are typically based on the loss of 50 percent of the materials mechanical properties. The selection of geomembrane polymer type is important to maximizing service life and should be factored into the designer's service life objective.

Conclusion

The Department has evaluated years of monitoring data substantiating that the performance of the double lined landfills are capable of containing the wastes that society cannot recycle or reuse without impacting groundwater quality. However, based on our experiences we have to acknowledge that these are not passive containment systems. Over time these systems tend to degrade and natural or manmade events can cause them to fail A significant level of maintenance is needed to ensure that the containment systems continue to function properly. This level of maintenance will decrease upon the end of operation at the facility, and will decrease again upon facility closure, but will not disappear completely. Containment system monitoring needs to be a key component for the post-closure period and some level of continued maintenance will always be necessary.

Our experience with older generation LLRW disposal sites and today's modern solid waste landfill has shown us that a modern LLRW disposal site will require ongoing monitoring and maintenance, and possible remedial actions when determined necessary to protect the public health and our groundwater resources as long as the waste materials remain a threat to public health and the environment.

Future endeavors should consider the following: we have realized a benefit exists from considering the use of interim containment designs for the short-term when faced with many long-term unknowns; we believe that more and continued research is needed on geosynthetic components used in containment systems relative to service life predictions; we should seek out and pursue benefits from emerging technologies such as nanotechnology developments/applications in geosynthetics for remote sensing, improved performance and durability; and, we know that we could enhance our containment system's long-term performance by incorporating reactive treatment barriers and remedial designs into the containment system design to mitigate potential future impacts.

References

1. NYSDEC (New York State Department of Environmental Conservation), 1992, *US EPA Region II Administrative Order on Consent Docket II,* RCRA 3008(h)-92-0202, Proceedings Under Section 3008(h) of the RCRA amended, March.

2. Final Environmental Impact Statement for Decommissioning and/or Long-Term Stewardship at the West Valley Demonstration Project and Western New York Nuclear Service Center, USDOE January 2010 http://www.westvalleyeis.com/finaleis.htm

3. Record of Decision: Final Environmental Impact Statement for Decommissioning and/or Long-Term Stewardship at the West Valley Demonstration Project and Western New York Nuclear Service Center, USDOE, April 2010 http://www.westvalleyeis.com/ROD.pdf

4. New York State Department of Environmental Conservation. 1996, Order on Consent, Index #A7-0333095-08.

5. New York State Department of Environmental Conservation, 2002, Record of Decision, Cornell University, Radiological Disposal Site, Lansing, Tompkins County, Site 7-55-002.

6. Nosko, V., Andrezal, T., Gregor, T., Ganier, P. 1996. "SENSOR Damage Detection Systems (DDS) – The Unique Geomembrane Testing Method." *Geosynthetics: Applications, Design and Construction. Proceedings of the First European Geosynthetics Conference, EuroGeo 1.* Rotterdam, Netherlands. 743-748.

Figure 1

How Well Are New York State's Double-Lined Landfill Designs Working ?

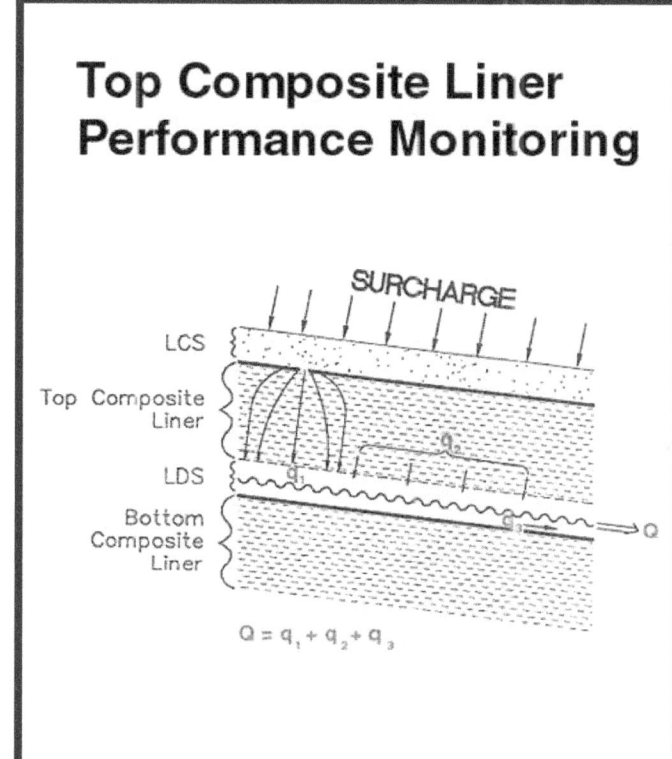

Top Composite Liner Performance Monitoring

From 2009 Annual Reports
(data on 31 Landfills)

Primary LCRS Flows:
Max: 9,249 gpad; Min: 233 gpad;
Mean: 1,281 gpad

Secondary LCRS Flows:
Max: 30.63 gpad; Min: 0.40 gpad;
Mean: 5.96 gpad

Upper Liner System Efficiency:
Max: 99.98 %; Min: 95.64 %;
Mean: 99.28 %

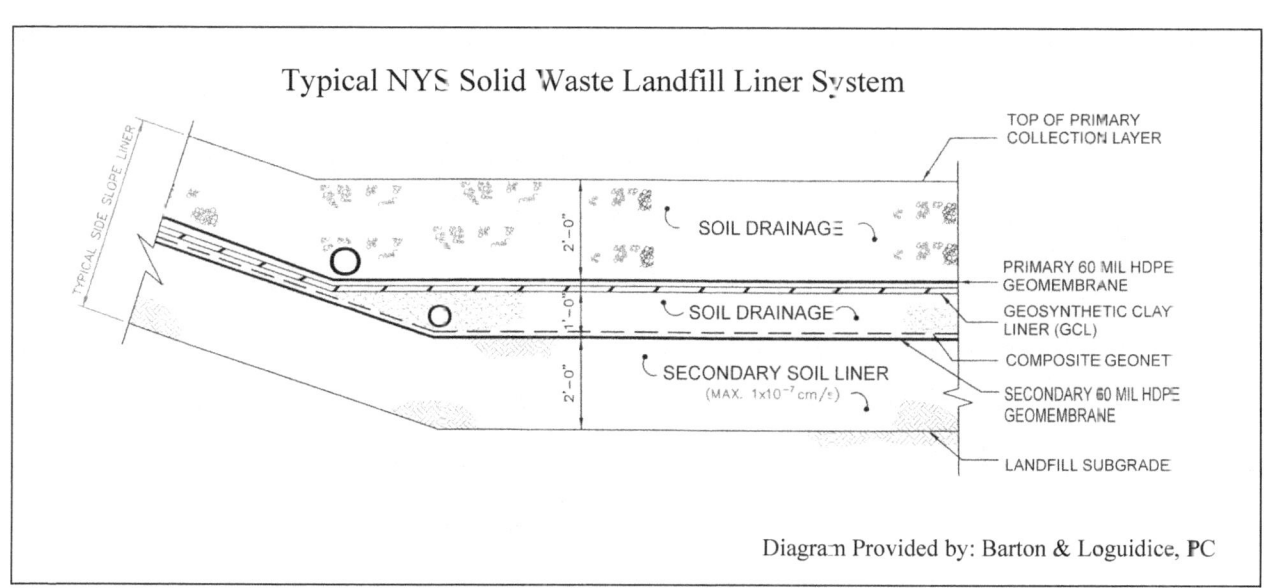

Typical NYS Solid Waste Landfill Liner System

Diagram Provided by: Barton & Loguidice, PC

SESSION 1B:

FEDERAL AGENCIES AND
DOE NATIONAL LABORATORIES

Session Chairs:
Jacob Philip, NRC/RES and
Brian Andraski, U.S. Geological Survey

Technical Reporter:
George Alexander, NRC/FSME

3.1 Motivation

Selected federal agencies and DOE national laboratories were requested to present and discuss their research activities and findings related to engineered cover performance which were employed to isolate wastes from the accessible environment. Their presentations focused on practical insights in modeling, monitoring and confirming short- and long-term performance of the engineered barriers systems. Although some federal agencies and DOE national laboratories conduct research on the performance of engineered waste covers under their purview, details on the linkage between iterative predictive performance assessment modeling and monitoring, and performance confirmation are generally not available. The purpose of this session was to present that information to a wider audience and to provide documentation on those details.

3.2 Background

The participants in this session were specifically requested to focus their presentations (and extended abstracts) in response to the following set of questions:

1. What performance assessment (PA) was done to predict dose due to gaseous and fluid releases from the waste facility?
2. What laboratory and field tests were performed to obtain input parameters for the PA modeling?
3. What field and laboratory testing was performed, and what measurements were taken to validate PA model results?
4. Is field monitoring continuing and at what time intervals to validate that the facility is continuing to perform to regulatory criteria?
5. What maintenance and repair activities are conducted to remediate the facility if regulatory criteria are not being met?
6. Are the PA's that were conducted for the sites and the laboratory/field test results publically available?

3.3 Challenges/Summaries of Session Presentations

A total of six presentations were made in this session. The presentations discussed the general theme of the session, modeling, monitoring and performance confirmation, in the context of actual research activities or, in one case, within the underlying regulatory context.

The U.S. Army Corps of Engineers - Omaha District, presentation described experiences with non-radiological hazardous waste landfills. The landfills consist of mixed wastes, municipal solid wastes, construction debris and other hazardous wastes. In the design phase of the landfills, representative samples of soil materials are collected and analyzed for obtaining design parameters for the design of the facility. Potential liquid releases through the proposed landfill are modeled using a hydrologic water-balance code such as HELP or UNSAT-H. During operation of the facility, leachate is collected and compared to predicted leachate quantities modeled in the pre-construction phase. Similarly, gas from the facility collection systems is analyzed and compared to the model predictions. Comprehensive reviews of the hazardous landfill facilities are performed every five years with repairs performed as needed. Reports from these programs are not publically available, but can be released to the public only via the FOIA process.

The EPA's Office of Radiation and Indoor Air (ORIA) presentation did not discuss activities connected to reviews of specific facilities. Rather, the presentation described EPA's review of its regulatory requirements for uranium and thorium Mill tailings codified in Chapter 40 of the Code of Federal Regulations, Part 192 (40 CFR Part 192). EPA is evaluating the performance history of Title 1 and Title 2 mill tailings sites against the standards in 40 CFR Part 192. The regulations are being reviewed due to changes in: uranium industry technologies and practices; potential inputs to ground water; recent revisions in EPA drinking water protection standards; judicial decisions concerning the regulations; minimization risks to the general public and environment, as well as, potential inputs to Tribes, environmental justice communities and children's health.

The Idaho National Laboratory (INL) in Idaho Falls, Idaho, presentation focused on providing practical insights on the modeling, monitoring and performance confirmation of natural and engineered systems. Input parameters for PA for one application were obtained by laboratory and field investigations. Determining the partitioning of carbon-14 (C-14) in the subsurface involved large-scale unsaturated column tests to assess the validity of PA models and its predictions. In another application, field measurements by advanced tensiometers were used to estimate bounds on variables used in the PA. INL field measurements have provided improved understanding for long-term monitoring in the vadose zone and deeper ground water systems, and have contributed to development of innovative instruments and techniques for use in various environments. As a consequence, these techniques have aided in the confirmation of long-term performance of waste facilities in the natural and engineered environment. Reports from these investigations are available and can be accessed by the scientific/engineering community at: https://inlportal.inl.gov/portal/server.pt/community/research_library/263/general_information .

The Savannah River Site E-Area presentation discussed the disposal of bulk and containerized LLW within shallow land disposal units. Due to the variable nature of the waste, the subsidence potential for the waste form can vary from 0 to 13.5 feet. Modeling of the engineered system to quantify potential subsidence of the waste form included an evaluation of the effects of static and dynamic compaction, and the time to total subsidence. Once the total subsidence has been reached, a final cover system should not be prone to further settlement and should provide a stable facility.

The Department of Energy's Office of Environmental Management (EM) presentation discussed DOE's perspective on PA modeling, monitoring and performance confirmation. The talk focused on information in DOE's Order 435.1- Chg 1 that describes the formal reviews of facility PA's. It includes specific requirements for comparing monitoring results with the PA (please see URL: https://www.directives.doe.gov/directives/current-directives/435.1-DManual-1c1/view). The reviews include probabilistic approaches for sensitivity and uncertainty analyses for PA s and performance monitoring of a broad range of indicators of changes in performance of the facility at specific locations and times based on PA.

The DOE Office of Legacy Management (LM) presentation discussed the long-term surveillance and maintenance of sites under UMTRCA and CERCLA. EPA's regulations in 40CFR 192 sets forth the standards for compliance with soil, air and ground water. Radon flux measurements are conducted during cell construction and upon cell completion. Post construction monitoring of radon is not required by the regulation. The disposal cells are designed using the code RAECOM to predict radon emanation from the cell. The final radon barrier thickness required is determined to limit the radon release to 20 picocuries/square meter/second. The RAECOM code requires field measurements and laboratory test data of the soils and the curie content of the waste. Ground-water protection strategies require ground-water monitoring and annual sampling at some sites at the point of compliance every year for 10 years and once every 5 years thereafter.

3.4   Recommendations

Panel discussions were held following the presentations. The panel members were composed of the presenters for the session and the session co-chairs, Jacob Philip and Brian Andraski. Significant observations and recommendations from the panel discussions were as follows:

- The need for monitoring adjacent to the waste containment facility. This aspect was deemed very important in light of 10 CFR 20.1406, whereby new nuclear facilities have to minimize contamination of the environment and the generation of waste. This implies the need for improved real-time detection of inadvertent release of contaminated materials from the facility.

- The importance of vadose zone monitoring. Vadose zone monitoring has been shown to be crucial for verifying the performance of waste facilities in both humid and arid environments. Application and further development of indirect (e.g., soil moisture) and direct (contaminant) monitoring strategies will strengthen PA activities.

- The importance of scale effects for input parameters in PA. Field measurements for both PA input parameters and performance monitoring of the facility are preferred.

- Continue using results from PA monitoring and modeling to identify and prioritize research needs.

3.5   Extended Abstracts

Following the review of technical literature in engineered barrier performance, the workshop organizing committee discussed significant ongoing projects within the Federal community. Invited speakers from Federal agencies and DOE national laboratories such EPA, U.S. Army Corps of Engineers, Pacific Northwest National Laboratories, Savannah River National Laboratory, and DOE contractors provided the following extended abstracts. These papers focus on practical insights in modeling, monitoring and confirming short- and long-term performance of the engineered barriers systems.

3.5.1 Waste Isolation and Contaminant Migration—Tools and Techniques for Monitoring the Saturated Zone-Unsaturated Zone-Plant-Atmosphere Continuum

Brian J. Andraski[1] and David A. Stonestrom[2]

U.S. Geological Survey
[1]Carson City, NV and [2]Menlo Park, CA

In 1976 the U.S. Geological Survey (USGS) began studies of unsaturated zone hydrology next to the Nation's first commercial disposal facility for low-level radioactive waste (LLRW) near Beatty, NV. Recognizing the need for long-term data collection, the USGS in 1983 established research management areas in the vicinity of the waste-burial facility through agreements with the Bureau of Land Management and the State of Nevada. Within this framework, the Amargosa Desert Research Site (ADRS; http://nevada.usgs.gov/adrs/) is serving as a field laboratory for the sustained study of water-, gas-, and contaminant-transport processes, and the development of models and methods to characterize flow and transport. The research is built on multiple lines of data that include: micrometeorology; evapotranspiration; plant metrics; soil and sediment properties; unsaturated-zone moisture, temperature, and gas composition; geology and geophysics; and groundwater. Contaminant data include tritium, radiocarbon, volatile-organic compounds (VOCs), and elemental mercury. Presented here is a summary of monitoring tools and techniques that are being applied in studies of waste isolation and contaminant migration.

Hydrologic performance of a waste-cover system can be strongly influenced by seasonal and annual variations in the amount of water delivered by precipitation and removed by evapotranspiration (ET). Multi-year monitoring of micrometeorological conditions and ET at the ADRS is capturing large temporal variations in climate. For example, annual precipitation averages 112 mm (1981–2005) and has ranged from a minimum of 3.5 mm (2002) to a maximum of 225 mm (1983). Details of the continuous micrometeorological and ET data collection systems are provided in Johnson and others (2007). Portable-chamber measurements also are being used to periodically measure localized ET and to partition it into bare-soil evaporation and plant transpiration components (Garcia and others, 2008; 2009).

Water moves through soil in liquid and vapor form, and the two forms move interdependently as a consequence of water potential and temperature gradients. Under dry conditions, vapor flow may dominate over liquid flow and several variables—water content, water potential, and temperature—are needed to define rates and directions of water movement. At the ADRS, soil water content is monitored using neutron-moisture probes, and water potential and temperature are monitored using thermocouple psychrometers (Andraski, 1997). Soil-water balance and water movement studies are done under both natural-site and simulated waste-site conditions. Monitoring results have identified vegetation as a main control on the presence or absence of deep percolation in desert regions (Gee and others, 1994). The data also have documented episodic, deep drying well below the root zone of native vegetation and established the pivotal role of plants in maintaining the upward movement of water in undisturbed desert unsaturated zones (Andraski, 1997). Long-term (>15 yr) monitoring shows that native plants remove virtually all infiltrated precipitation on an annual basis. Under nonvegetated waste-site and test-trench conditions, data show the accumulation and continued (albeit slow) penetration of precipitation in the uppermost 2.75 m of the profiles (Johnson and others, 2007). Wind erosion and subsidence of the test trenches also are being monitored to evaluate factors that can affect infiltration and the long-term hydrologic performance of trench covers. These measurements are made by surveying a series of reference pins and plates (Andraski, 1997; Andraski and Prudic, 1997).

The test-trench soil-moisture monitoring techniques used at the ADRS have been applied to the hydrologic performance assessment of a water-balance cover at Fort Carson, CO. In that work, multiple-year soil-moisture and lysimeter measurements, and calculations of downward water fluxes provided complementary data and results to support the conclusion of negligible annual drainage over the 5-yr test period (McGuire, Andraski, and Archibald, 2009). Soil-water potential data showed that ET not only removed infiltrated water, but also contributed to a persistent driving force for removal of water from below the base of the cover.

ADRS studies of hydrologic processes in the thick unsaturated zone use data from instrumented boreholes extending from land surface to the 110-m deep water table. Data show that natural gradients of water potential, temperature, and carbon dioxide provide for upward movement of water and gas through the unsaturated zone beneath undisturbed native vegetation (Andraski and Prudic, 1997; Stonestrom and others, 1999; 2004; Walvoord and others, 2005). Techniques for deep-borehole drilling, installation of water-potential and temperature sensors, and instrumentation and sampling of soil-gas ports are described by Prudic and Striegl (1995), Andraski and Scanlon (2002), and Walvoord and others (2005).

A study to determine the natural distribution of gases and depth of atmospheric-air circulation in the unsaturated zone uncovered the presence of radioactive contaminants moving laterally beneath the ADRS—160 m from the nearest LLRW disposal trench (Prudic and Striegl, 1995). The LLRW, which contains both radioactive and hazardous components, was buried in unlined trenches (2-15 m deep) from 1962-1992. Contaminants must pass through the thick unsaturated zone before reaching the underlying aquifer, which remains relatively uncontaminated. Thus, the primary ADRS research focus is on processes affecting contaminant migration in the unsaturated zone. Data on contaminant migration comes from an extensive array of shallow gas-sampling tubes (<2 m deep), two instrumented boreholes (up to 109 m deep), air and plant sampling sites, and groundwater sampling wells.

In strong contrast to the natural upward-directed flow field of water, gas, and heat, the contaminant monitoring data show that tritium, radioactive-carbon dioxide, VOCs, and elemental mercury are migrating laterally from the waste source. The ADRS multiple-contaminant sampling approach in combination with knowledge of hydrogeologic conditions indicate the long-distance (>300 m) lateral migration is predominantly gas-phase transport that occurs preferentially through porous gravelly layers. Contaminant sampling includes various gas-pumping and trapping techniques. Tritium sampling of air and unsaturated-zone gas is done by condensing water vapor in a dry-ice trap and analyzing the resultant liquid by scintillation counting (Prudic and Striegl, 1995; Andraski and others, 2003). Radioactive-carbon dioxide is collected in sodium-hydroxide solutions, precipitated as barium carbonate, and measured by liquid scintillation counting (Stonestrom and others, 2004). VOCs are trapped on multisorbent columns and quantified by mass spectroscopy (Pankow and others, 1998). Elemental mercury is captured on gold traps and measured by cold-vapor atomic fluorescence spectrometry (Walvoord and others, 2008).

Extracting a single soil-water vapor sample for tritium analyses using traditional techniques often requires >12 hours of pumping soil gas. A new method developed at the ADRS uses native shrubs and solar-distillation of foliage to collect unsaturated zone water for tritium analysis (Andraski and others, 2003; 2005). Field sampling is rapid—in two days, sufficient plant samples can be collected and distilled to delineate a kilometer-sized plume. The method is noninvasive and, since plant-derived water samples come from active roots, it provides a volume-integrated sample of the subsurface that is much larger than that for tube-based samplers. In a recent ADRS study, plant- and soil-tritium data were combined with ET measurements to quantify the plume-scale flux of tritium from the subsurface to the atmosphere (Garcia and others, 2009).

Plant-based mapping can delineate contamination of shallow sediments in considerable detail, but delineating contamination in deep unsaturated sediments is hindered by the limited number of multilevel-sampling boreholes. To extend measurements made with a small number of deep boreholes, recently developed geophysical techniques permit the imaging of the subsurface strata that strongly control contaminant transport. Noninvasive, direct-current resistivity measurements show high resistivity layers at depths that correspond to the maximum detected levels of the principal gaseous contaminants (Stonestrom and others, 2004; 2007; Lucius and others, 2008). These gravelly layers are continuous over large horizontal distances and serve as preferential paths for migration of contaminants from the mixed-waste source.

References

Andraski, B.J., 1997, Soil-water movement under natural-site and waste-site conditions: A multiple-year field study in the Mojave Desert, Nevada. Water Resources Research: v. 33, p. 1901–1916, doi:10.1029/97WR01502.

Andraski, B.J., and Prudic, D.E., 1997, Soil, plant, and structural considerations for surface barriers in an arid environment—Application of results from studies in the Mojave Desert, in Barriers for Environmental Management, Summary of a Workshop: Washington, D.C., National Academy Press, p. D50-D60.

Andraski, B.J., and Scanlon, B.R., 2002, Thermocouple psychrometry, in Dane, J.H., and Topp, G.C., eds., Methods of Soil Analysis, Part 4, Physical Methods: Madison, WI, Soil Science Society of America, p. 609-642.

Andraski, B.J., Sandstrom, M.W., Michel, R.L., Radyk, J.C., Stonestrom, D.A., Johnson, M.J., and Mayers, C.J., 2003, Simplified method for detecting tritium contamination in plants and soil: Journal of Environmental Quality, v. 32, no. 3, p. 988-995, doi:10.2134/jeq2003.0988.

Andraski, B.J., Stonestrom, D.A., Michel, R.L., Halford, K.J., and Radyk, J.C., 2005, Plant-based plume-scale mapping of tritium contamination in desert soils: Vadose Zone Journal, v. 4, p. 819-827, doi:10.2136/vzj2005.0052.

Garcia, C.A., Andraski, B.J., Stonestrom, D.A., Cooper, C.A., Johnson, M.J., Michel, R.L., and Wheatcraft, S.W., 2009, Transport of tritium contamination to the atmosphere in an arid environment Vadose Zone Journal, v. 8, no. 2, p. 450-461, doi: 10.2136/vzj2008.0022.

Garcia, C.A. Johnson, M.J., Andraski, B.J., Halford, K.J., and Mayers, C.J., 2008, Portable chamber measurements of evapotranspiration at the Amargosa Desert Research Site near Beatty, Nye County, Nevada, 2003–06: U.S. Geological Survey Scientific Investigations Report 2008-5135, 10 p. (Available at http://pubs.usgs.gov/sir/2008/5135/)

Gee, G.W., Wierenga, P.J., Andraski, B.J., Young, M.H., Fayer, M.J., and Rockhold, M.L., 1994, Variations in water balance and recharge potential at three western desert sites: Soil Science Society of America Journal, v. 58, no. 1, p. 63-72.

Johnson, M.J., Mayers, C.J., Garcia, C.A., and Andraski, B.J., 2007, Selected micrometeorological, soil-moisture, and evapotranspiration data at Amargosa Desert Research Site in Nye County near Beatty, Nevada, 2001–05: U.S. Geological Survey Data Series 284, 28 p. (Available at http://pubs.usgs.gov/ds/2007/284/)

Lucius, J.E., Abraham, J.D., and Burton, B.L., 2008, Resistivity profiling for mapping gravel layers that may control contaminant migration at the Amargosa Desert Research Site, Nevada: U.S. Geological Survey Scientific Investigations Report 2008–5091, 30 p. (Available at http://pubs.usgs.gov/sir/2008/5091/)

McGuire, P.E., Andraski, B.J., and Archibald, Ryan, 2009, Case study of a full-scale evapotranspiration cover: Journal of Geotechnical and Geoenvironmental Engineering, v. 135, no. 3, p. 316-332, doi:10.1061/(ACSE)1090-0241(2009)135:3.

Pankow, J.F., Luo, Wentai, Isabelle, L.M., Bender, D.A., and Baker, R.J., 1998, Determination of a wide range of volatile organic compounds in ambient air using multisorbent adsorption/thermal desorption and gas chromatography/mass spectrometry: Analytical Chemistry, v. 70, no. 24, p. 5213-5221.

Prudic, D.E., and Striegl, R.G., 1995, Tritium and radioactive carbon (14C) analyses of gas collected from unsaturated sediments next to a low-level radioactive-waste burial site south of Beatty, Nevada, April 1994 and July 1995: U.S. Geological Survey Open-File Report 95-741, 7 p.

Stonestrom, D.A., Abraham, J.D., Andraski, B.J., Baker, R.J., Mayers, C.J., Michel, R.L., Prudic, D.E., Striegl, R.G., and Walvoord, M.A., 2004, Monitoring radionuclide contamination in the unsaturated zone--Lessons learned at the Amargosa Desert Research Site, Nye County, Nevada: Proceedings, Workshop on long-term performance monitoring of metals and radionuclides in the subsurface, Reston, VA, April 20-22.

Stonestrom, D.A., Prudic, D.E., and Striegl, R.G., 1999, Isotopic composition of water in a deep unsaturated zone beside a radioactive-waste disposal area near Beatty, Nevada, in Morganwalp, D.W., and Buxton, H.T., eds., U.S. Geological Survey Toxic Substances Hydrology Program--Proceedings of the Technical Meeting, Charleston, South Carolina, March 8-12, 1999--Volume 3 of 3--Subsurface Contamination from Point Sources: U.S. Geological Survey Water-Resources Investigations Report 99-4018-C, p. 467-474.

Stonestrom, D.A., Prudic, D.E., Walvoord, M.A., Abraham, J.D., Stewart-Deaker, A.E., Glancy, P.A., Constantz, J., Laczniak, R.J., and Andraski, B.J., 2007, Focused ground-water recharge in the Amargosa Desert Basin, in Stonestrom, D.A., Constantz, J., Ferré, T.P.A., and Leake, S.A., eds., Ground-water recharge in the arid and semi-arid southwestern United States: U.S. Geological Survey Professional Paper 1703, p. 107-136.

Walvoord, M.A., Andraski, B.J., Krabbenhoft, D.P., and Striegl, R.G., 2008, Transport of elemental mercury in the unsaturated zone from a waste disposal site in an arid region: Applied Geochemistry, v. 23, no. 3, p. 572–583, doi:10.1016/j.apgeochem.2007.12.014.

Walvoord, M.A., Striegl, R.G., Prudic, D.E., and Stonestrom, D.A., 2005, CO_2 dynamics in the Amargosa Desert–fluxes and isotopic speciation in a deep unsaturated zone: Water Resources Research, v. 41, doi:10.1029/2004WR003599.

3.5.2   USACE Experience with HTW Containment Systems

Kevin Pavlik

U.S. Army Corps of Engineers

USACE-Omaha District's landfill containment experience is primarily with non-radiological hazardous waste repositories. The majority of landfill design experience is with military landfill cover systems. Military landfills typically consist of mixed wastes; municipal solid waste, construction debris, and hazardous wastes, but not radiological wastes.

Production rates for gas collection system design are estimated based on waste volume and type or quantitatively field measured. Landfill gas consists largely of methane and carbon dioxide, which is often passively vented to the atmosphere or actively collected and flared.

Liquid releases through cover and liner systems are calculated via hydrologic water balance modeling. One such computer model is the Hydrologic Evaluation of Landfill Performance (HELP). UnSat-H is another program available for hydrologic water balance analyses. Typical hydrologic modeling input parameters consist of climate, vegetation, slope, material properties and thicknesses.

Prior to construction, representative material samples are collected and analyzed for consistency with the design calculations and assumptions. Typical geotechnical testing consists of grain size distribution, plasticity (Atterberg limits), moisture-density relationships, and hydrologic conductivity. Layer thicknesses are monitored using traditional topographic survey methods.

During operation, leachate collected in lined systems can be compared to predicted volumes and constituents and to determine the systems effectiveness. Leachate system effectiveness is monitored by sampling from ground water wells and testing for analytes of concern. Similarly, the effectiveness of the gas collection system can be monitored by sampling from wells above the ground water and testing for analytes of concern.

Comprehensive reviews of hazardous waste landfills are performed every five-years. Operations and Maintenance manuals typically require cover systems be mowed annually and inspected after major precipitation or earthquake events.

Repair activities are typically required due to erosion of the cover materials or veneer slope failure. Repair activities typically consist of reestablishment of the deficient cover material types/thickness and vegetation.

Project data is provided as needed to parties involved with the project, but is only released to the public when requested via the Freedom of Information Act (FOIA).

3.5.3   Design Basis of UMTRCA Disposal Cells

Rich Bush, U.S. Department of Energy, Office of
Legacy Management, Grand Junction, Colorado
John Elmer, S.M. Stoller Corporation, Grand Junction, Colorado
Greg Smith, Geo-Smith Engineering, LLC, Grand Junction, Colorado

The DOE Office of Legacy Management (LM) currently performs long-term surveillance and maintenance of sites with uranium mill tailings disposal cells at 30 locations across the country. Most of the Uranium Mill Tailings Radiation Control Act (UMTRCA) cells are in the western United States. Other disposal cell sites are under CERCLA regulation and may have design parameters and requirements that are different from those of UMTRCA disposal cells.

Congress enacted UMTRCA in 1978 after recognizing that active and inactive uranium mill tailings sites posed a threat to human health and the environment. The risks were primarily from radon emanation and physical access to exposed tailings. UMTRCA was designed to address this health threat by stabilizing the tailings and limiting radon emanation. The regulation came shortly after enactment of RCRA, which was created to address hazardous wastes at active sites, and CERCLA, which addressed hazardous wastes at inactive, abandoned sites.

The surface remediation program under UMTRCA had an initial term of 10 years for DOE to complete cleanup at designated sites. This time frame proved overly optimistic, and after several amendments, the authority for surface cleanup expired in 1998. Under Title I of UMTRCA, DOE performed the cleanup of inactive mill sites, and under Title II, the licensees of active mill sites continued to remediate their sites under license from the NRC or agreement states. After completion of cleanup, DOE assumes long-term custody of both Title 1 and Title II sites.

Disposal cell design and remediation goals for the Title I sites evolved over the 20-year surface program. The courts remanded the initial groundwater standards during the early stages of cleanup, and the revised standards issued in 1987 represented a change to more conservative concepts at the sites where disposal cell construction had not yet begun. The final standards were not issued until 1995, well after most cells were designed.

EPA regulations in 40 CFR 192, "Health and Environmental Protection Standards for Uranium and Thorium Mill Tailings," established maximum concentration limits for residual radioactive materials in soil, air, and groundwater. Requirements in 40 CFR 192 included the use of approved designs of disposal cells to control radon flux and also included the development of groundwater protection plans through site characterization and monitoring.

The regulations do not require post-construction monitoring of disposal cell performance (40 CFR 192.02 [a], [b]). The assumption is that the design for isolation of residual radioactive materials would be adequate if design parameters were met during construction. However, groundwater monitoring is required at points of compliance to verify success of the groundwater protection plan (40 CFR 192.02 [c]).

To provide adequate protection from radon emanation and to provide a design basis for the disposal cells, the computer code RAECOM is used to predict radon (Rn-222) release from a cell based on design parameter input. During tailings emplacement in the disposal cell, lower-activity materials were typically placed over higher-activity materials to decrease the radon emissions and decrease the required thickness of the overlying radon barrier. The final required radon barrier thickness is designed to limit the radon release to less than 20 picocuries per square meter per second, the EPA standard (40 CFR 192.02 [c]).

The RAECOM code requires input of field measurements, laboratory test data, and several design parameters. Input data include the disposal cell layer thickness, which is a design parameter; dry bulk density obtained from field measurements; porosity calculated from the bulk density; long-term moisture content as measured in the field; a radon emanation coefficient and radon diffusion coefficient from lab testing and finally, the radium-226 concentration of the tailings, as measured in the laboratory.

Rn-222 concentrations are measured at the site before construction to establish background levels and during construction using alpha-track detectors located at the site boundaries. During cell construction and after completion of the radon barrier, radon flux measurements are taken on the surface using large-area activated charcoal canisters in conformance with EPA requirements to verify the RAECOM results.

No post-construction radiological performance assessment is required or routinely performed; however, radon measurements have been taken at the Lakeview, Oregon, and the Shiprock, New Mexico, sites. These measurements indicate that radon releases at both sites are well within performance standards.

A groundwater protection plan at a minimum must contain a list of hazardous constituents likely to be present and derived from residual radioactive materials at the site, as well as maximum concentration limits or alternate concentration limits, as appropriate. The plan is based on the 1996 *Final Programmatic Environmental Impact Statement for the Uranium Mill Tailings Remedial Action Ground Water Project* (PEIS). The PEIS provides a framework from which to apply a groundwater compliance strategy that will comply with EPA standards in 40 CFR 192.03. The protection plan includes a description of the overall protection strategy, a description of the strategy for compliance for each constituent, and a description of the points of compliance for the uppermost aquifer at the disposal site.

The groundwater monitoring required at the Gunnison and Rifle, Colorado, disposal sites illustrate these requirements. At the Gunnison site, a multicomponent disposal cell cover limits infiltration of moisture. Contaminated materials were emplaced at low moisture content to limit transient drainage. Favorable geochemical conditions in the foundation soils also attenuate contaminant movement, and the tailings and other contaminated materials were removed from the mill site to the current location to take advantage of these conditions. The uppermost aquifer is 100 feet below the disposal cell, and the nearest downgradient user is approximately 7,500 feet from the site. The monitoring plan requires annual sampling of the point-of-compliance well for 10 years and once every 5 years thereafter to provide continuing assurance of protectiveness.

The Rifle disposal site also includes a multicomponent disposal cell cover to limit moisture infiltration. Transient drainage was modeled prior to cell construction and was to some extent expected, and a plan to address the drainage was included in the Long Term Surveillance Plan. This plan included placement of standpipes in the cell and a requirement to remove excess tailings pore water above a specified action level. That action level was reached after several years and prompted the construction of a lined evaporation pond to store and evaporate fluid pumped from the toe of the cell. The disposal cell site was chosen in a geologically favorable location where the uppermost aquifer is several hundred feet below the cell and is classified as limited use based on low yield (less than 150 gallons per day) and high total dissolved solids (over 10,000 milligrams per liter). In addition, widespread ambient contamination unrelated to ore processing exists at this site, and the groundwater cannot be cleaned up using treatment methods currently employed in public water supply systems; therefore, no monitoring is required (40 CFR 192.11 [e] [1]).

A recent study at the Lakeview disposal site is also relevant. Previous work indicated the presence of saturated tailings, which prompted further investigation. The concern over this condition was mainly that the seismic stability of the disposal cell may be in question if saturation was consistent throughout the cell. A plan was devised to use a Geoprobe to penetrate the tailings at various depths along two transects on the cell to determine the extent of saturated tailings. The investigation found evidence of perched water in tailings in one hole at about 16 feet below the surface, which might be an indication of transient cell drainage. Following this investigation, all boreholes were sealed. Only very low levels of radiation were detected, even in the open tailings portions, possibly due to the preferential placement of low-activity tailings on top of higher-activity materials. As a result of this study, concern over seismic stability is greatly reduced.

As part of its ongoing mission to protect human health and the environment, LM performs routine inspections and maintenance as required by the NRC-approved Long-Term Surveillance Plans, which serve as our licensing documentation. LM has collected data on disposal cells for more than 20 years during various investigations that have included inspections, groundwater monitoring, and limited radon monitoring. Through the Applied Science and Technology program, LM periodically performs studies to investigate potential concerns, optimize the performance and understanding of the sites, and develop possible solutions before any significant threats to human health or the environment are identified. Studies have included an investigation of disposal cell cover performance and the possibility of performing some sort of cover renovation if necessary. This study is ongoing and will take some time to provide usable data and methods to enhance cover performance. Thus, it is important to begin these types of studies prior to the need. Once a practical methodology for cover renovation is developed at the Grand Junction disposal cell, application of the technology at another UMTRCA disposal site would require a site-specific study to tailor the renovation to the site. Thus, several years would pass between the time the need for a cover renovation is identified and a solution could be applied.

The Grand Junction disposal cell is close to the LM office in Grand Junction and has proven to be an ideal location to test methods of renovating disposal cell covers. The site has ample room, a readily available source of cover material identical to that already used at the disposal cell, and a climate similar to that of a number of other sites with UMTRCA cells.

The main goal of this study is to develop a method to renovate a disposal cell cover by converting it from one that was designed as a simple, engineered rock cover to one that can sustain evapotranspiration, a feature that is also discussed during the workshop. Reasons to consider evapotranspiration as an option include being to eliminate the current requirement to cut and spray undesirable plants on the cell, and enhancing the removal of water from the cell to minimize infiltration. A need for this method has not yet been identified at any of the LM disposal cell sites, but because of the lengthy time required to gather data and develop the renovation methodology, it was deemed prudent to begin the studies now before these technologies might be needed.

LM is performing long-term surveillance and maintenance at 30 disposal cell sites across the country under UMTRCA and CERCLA regulations, in varying climates, and with different containment requirements. Performance at all LM sites remains within NRC requirements and EPA standards and is protective of human health and the environment. LM continues to develop new technologies and process improvements to maintain protection of human health and the environment at an acceptable cost to the taxpayer, and to provide timely responses to any problems that occur.

References

DOE (U.S. Department of Energy) 1994. *Long-Term Surveillance Plan for the Collins Ranch Disposal Site, Lakeview, Oregon,* DOE/AL/62350-19F Rev. 3, August.

DOE (U.S. Department of Energy) 1994. *Long-Term Surveillance Plan for the Shiprock Disposal Site Shiprock New Mexico,* DOE/AL/62350-60F Rev.1, September.

DOE (U.S. Department of Energy) 1996. *Final Programmatic Environmental Impact Statement for the Uranium Mill Tailings Remedial Action Ground Water Project,* DOE/EIS-0198, October.

DOE (U.S. Department of Energy) 1997. *Long-Term Surveillance Plan for Gunnison, Colorado Disposal Site,* DOE/AL/62350-222 Rev.2, April.

DOE (U.S. Department of Energy) 1997. *Long-Term Surveillance Plan for the Estes Gulch Disposal Site Near Rifle, Colorado,* DOE/AL/62350-235 Rev.1, November.

DOE (U.S. Department of Energy) 2009. *Technical Approach Document,* UMTRA-DOE/AL 050425.0002 Rev. 1, December.

3.5.4   EPA'S Review of its Regulatory Requirements for Uranium and Thorium Mill Tailings: 40 CFR Part 192

Loren W. Setlow

U.S. Environmental Protection Agency
Office of Radiation and Indoor Air (6608J)
Washington, DC 20460
Setlow.loren@epa.gov

The U.S. Environmental Protection Agency (EPA) is currently reviewing and may revise its existing regulations in 40 CFR Part 192, written under authority of the Uranium Mill Tailings and Radiation Control Act (UMTRCA). These regulations establish health, environmental protection and safety standards to be utilized by the U.S. Department of Energy (DOE), as well as the U.S. Nuclear Regulatory Commission (NRC) and its Agreement States for overseeing uranium and thorium mill tailings facilities. Section 275(b)(2) of the Atomic Energy Act (AEA) (42 U.S.C. § 2022(b)(2)) authorizes the Administrator to periodically revise the standards. The regulations were first promulgated in 1983 and last revised in 1995, but they are being reviewed due to changes in uranium industry technologies and practices, potential impacts to groundwater, recent revisions in EPA drinking water protection standards, judicial decisions concerning the subject regulations, unanticipated risks to the general public and environment as well as potential impacts to Tribes, environmental justice communities, and children's health. As part of its effort, the Agency will be reviewing information on the performance history and protectiveness of conventional mill tailings covers and impoundment designs.

Section 206 of UMTRCA (AEA section 275) includes EPA standard setting provisions with respect to specific inactive (and abandoned) uranium mills now referred to as "Title I" mills and depository sites, and for other facilities which were in operation or might be licensed in the future by the NRC which produce (AEA section 11e.(2)) byproduct material, now referred to as "Title II" mills. The standards must be consistent with the requirements of the Solid Waste Disposal Act (SWDA), as amended (now the Resource Conservation and Recovery Act (RCRA)) for non-radiological hazards. However, the statute allows the Agency discretion in setting standards for controlling radiological hazards. The EPA regulations established design, radon, radioactivity, monitoring, groundwater protection and corrective action standards for Title I mills to be utilized by the DOE; reclamation standards for contaminated soil and buildings for use of DOE and NRC (as well as its Agreement States); plus design, operation and closure, radon, radioactivity, monitoring, groundwater protection and corrective action standards for operating and future uranium and thorium mills licensed by the NRC (and its Agreement States). Many of the standards cross-reference EPA RCRA requirements for Title C hazardous waste facilities.

In conducting its review, an EPA work group is evaluating the environmental performance history of Title I and Title II mills against the standards in 40 CFR Part 192. Agency participation at workshops such as this provide an opportunity to learn of advances in barrier and containment cell designs for effectively limiting radon emissions, as well as resisting erosion and preventing water contamination.

3.5.5   Investigations Supporting Performance Verification of Engineered Barrier Systems

Joel Hubbell, Jeff Sondrup, Art Rood

Idaho National Laboratory, Idaho Falls, ID

The Idaho National Laboratory has conducted research activities that provide practical insight on monitoring, modeling and performance of natural and engineered systems in challenging geologic environments. These investigations have provided information used to enhance performance assessments by understanding water and contaminant flow and transport in the vadose and groundwater from previous waste disposals practices. We will present information from several investigations that provide insight to short- and long-term performance of natural and engineered systems. The INL has conducted numerous investigations for sites with both fractured rock and sedimentary deposits with deep vadose zones and a highly transmissive aquifer. Input parameters for performance assessments often require estimation of field parameters based on tools and techniques that had to be developed to characterize these facilities. Examples of several laboratory and field investigations will be presented showing both positive outcomes and lessons learned. The first is a large scale unsaturated column test conducted to evaluate the partitioning of carbon-14 in the subsurface, determine parameters for migration and demonstrate the validity of models and predictions. This investigation indicated that a large portion of the C-14 would partition into the gas phase and be released into the atmosphere rather than migrate to the aquifer. In another investigation, advanced tensiometers were monitored in the deep subsurface to bound variables that were used for performance assessment calculations, characterize ambient moisture/contaminant flow conditions, define temporal recharge on the decadal scale, and verify that tensiometer monitoring can be used to confirm barrier covers are effective in decreasing long term vertical water flux. Another investigation involved a large scale infiltration/pumping test to define parameters for flooding scenarios in fractured rock intercalated with sedimentary interbeds while providing large scale aquifer flow characteristics that had not been measured previously. Follow on work developed a measurement technique to obtain precise water level measurements unencumbered by barometric influences. Other INL field trials have provided improved understanding for long-term monitoring of both ground water and vadose zones in humid to arid environments while assisting in developing instrumentation techniques and tools for monitoring in many challenging environments. These national laboratory developed advancements have allowed a better understanding of key elements of flow and transport in the subsurface that can be applied to monitoring, modeling and confirming short and long-term performance in natural and engineered environments.

3.5.6 Savannah River Site E-Area Low-Level Waste Facility Subsidence Studies SRNL-MS-2010-00128, Revision 1

Mark Phifer

Savannah River National Laboratory
Savannah River Site
Aiken, SC 29808

Introduction

Performance Assessment (PA) model support associated with implementation of cover system engineered barrier long-term performance (i.e. degradation) should focus initially on the primary degradation mechanisms. That is those degradation mechanisms which could have an overriding impact on infiltration through the cover system. Subsidence, in particular differential subsidence, is one such potential primary degradation mechanism, which should be evaluated for its relevance and impact on cover system engineered barrier long-term performance. The E-Area Low-Level Waste Facility (ELLWF) at the Savannah River Site (SRS) has subsidence concerns, which are being addressed as outlined herein.

The ELLWF is the SRS low-level radioactive waste disposal and storage facility that manages Low-Level Waste (LLW) resulting from SRS operations. The ELLWF manages the disposal and storage of both containerized and bulk (non-containerized) LLW. The ELLWF is a 100 acre facility that contains shallow land disposal units (Slit and Engineered Trenches), a grout-encapsulation unit (Component-in-Grout Trenches), concrete vaults (Low Activity Waste and Intermediate Level Vaults), and metal casks (Naval Reactor Component Disposal Areas). Waste Acceptance Criteria (WAC) has been developed for each disposal unit type that specifies the waste acceptable for disposal in each type unit with the level of containment commensurate with the activity of the waste. Figure 1 provides the ELLWF layout. (Phifer et al. 2009)

Ninety cubic foot carbon steel boxes constitute the primary container disposed in the shallow land disposal units at the ELLWF. These containers are utilized primarily to promote worker radiation protection during the generation, transportation, and disposal of SRS LLW. The waste contained within the boxes is generally a low-density waste. The subsidence potential of the ELLWF shallow land disposal units is highly variable due to waste variability. The subsidence potential could range from near zero for bulk waste to 13.5 feet for the typical stack of four boxes containing low-density waste (i.e. 16 foot waste zone). (Phifer and Wilhite 2001; Phifer et al. 2009) The subsidence potential of the ELLWF shallow land disposal units is a major area of Savannah River National Laboratory (SRNL) research and development (R&D) relative to the long-term performance of cover system engineered barriers.

SRNL has conducted R&D related to subsidence for over 20 years. This work has included quantification of the subsidence potential of ELLWF shallow land disposal units, effectiveness of subsidence treatment methods (static surcharge and dynamic compaction), and determination of the optimal subsidence treatment method and its timing. Additionally SRNL has contributed to the closure planning for the ELLWF shallow land disposal units to ensure that the subsidence potential is appropriately accommodated.

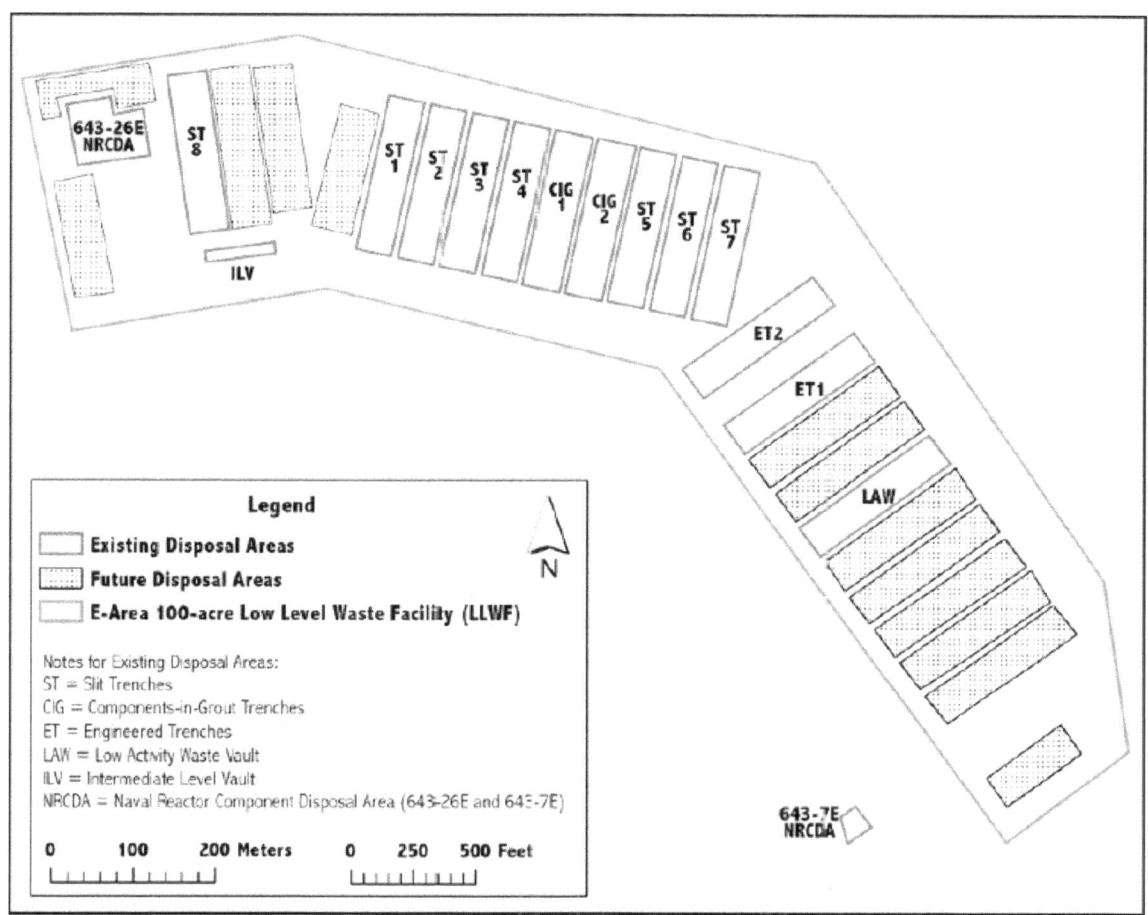

Figure 1. E-Area Low-Level Waste Facility (ELLWF) Layout

SRS Static Surcharge Test Program

A field evaluation of static surcharge for the stabilization of a 1.5-acre shallow land disposal unit containing typical stack of four carbon steel boxes was conducted during 1988 and 1989. The static surcharge field test consisted of the placement of subsidence monitors and a 25-foot soil surcharge on top of the trench, which had an interim soil cover from four to eight feet thick. Over a year period an average induced subsidence of 2.7 feet was measured over the north two thirds of the trench, which had had an average 7-foot interim soil cover. However a large percentage of the subsidence was due to consolidation within the 7-foot interim soil cover. It is estimated that consolidation within the 7-foot interim soil cover was approximately 1.3 feet. This leaves approximately 1.4 feet of induced subsidence that was due to box buckling. (Phifer, 1991; Phifer and Wilhite 2001)

SRS Dynamic Compaction Test Programs

Dynamic compaction consists of dropping a weight (up to 35 tons and up to 8 feet in diameter) from a specified height (up to 100 feet) repetitively in a specific pattern to compact soil and/or waste. Dynamic compaction depth of influence and overall compaction effectiveness increases with greater applied energy and a reduction in the displacement between drops (i.e. dropping a larger weight from a greater height until a minimum displacement (a fraction of a foot) between drops is achieved). Two primary dynamic compaction test programs have been conducted at SRS, one in 1988 and the other in 1993. (Phifer 1991; Phifer and Serrato 2000)

In 1988 a dynamic compaction test program was conducted on actual SRS Mixed Waste Management Facility (MWMF) shallow land disposal units containing non-containerized LLW. This program was conducted in order to evaluate the effectiveness and safety (primarily the potential for atmospheric releases and impact on surrounding structures) of dynamic compaction as a means of waste consolidation and to determine production specifications for closure of the MWMF. This program determined that dynamic compaction could be safely performed at the MWMF and could result in an estimated 11 to 18% waste volume reduction of the non-containerized LLW. (Phifer 1991)

In 1993 a dynamic compaction test program was conducted on the typical stack of four carbon steel boxes containing simulated waste in a 1.5-acre test shallow land disposal unit. This test program was conducted to support closure of the SRS Low-Level Radioactive Waste Disposal Facility (LLRWDF). This program evaluated the effectiveness of tertiary dynamic compaction treatment and the impact of dynamic compaction on adjacent compacted clay caps and determined production specifications for closure of the LLRWDF. This field evaluation concluded that in general the top boxes of the stack were more fully compacted than the boxes on the bottom. It was also observed that the boxes on the top layer were fused together by their lateral spread and interlocking, which may have inhibited the further effectiveness of dynamic compaction. Some boxes were actually breached (see Figure 2). Finally it was observed that the potential for box corrosion was increased due to the breakage of the protective coating bond with the metal and subsequent exposure of the bare metal. This evaluation determined that effectiveness could be increased by compacting in a pattern that completely covered the entire surface area and compacting until final displacements between drops were less than 0.2 feet. Such a tertiary compaction pattern and displacement criteria resulted in a 30% increase in compaction over the standard compaction criteria used to compact the 1.5-acre test trench. (McMullin and Dendler, 1994; Phifer and Serrato, 2000; Phifer and Wilhite 2001)

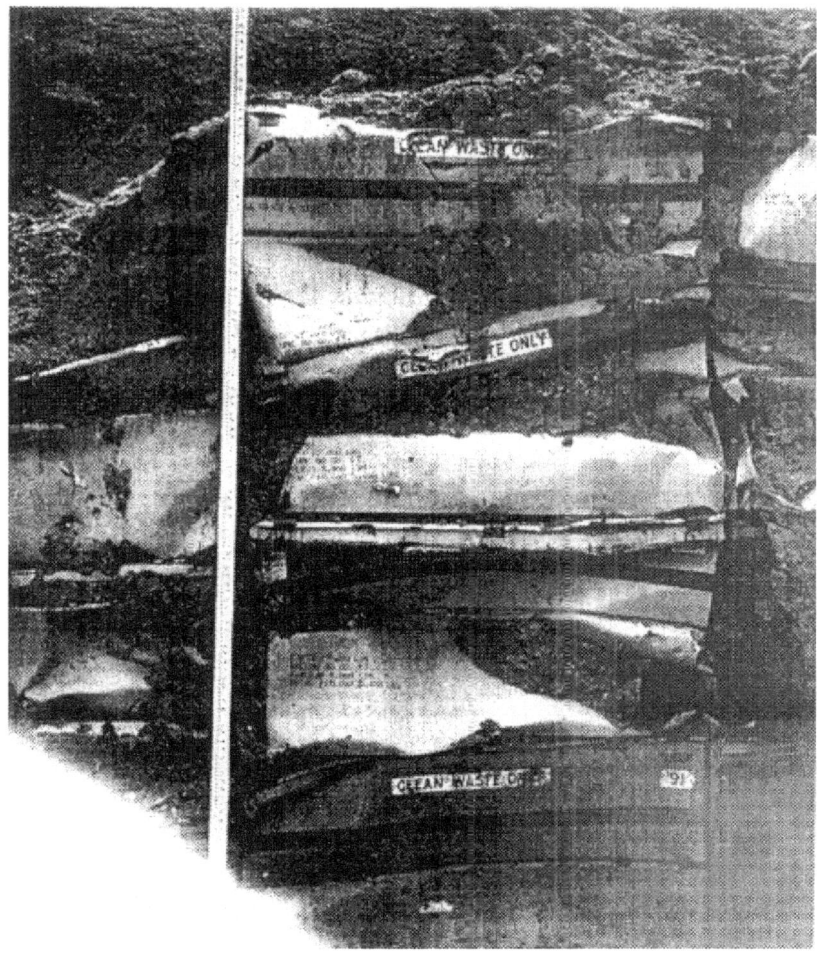

Figure 2. Dynamic Compaction of a Typical Stack of Four Carbon Steel Boxes

SRS Dynamic Compaction Projects

In 1989, standard dynamic compaction of 58 acres of the MWMF was conducted including the compaction of a 1.5-acre shallow land disposal unit containing the typical stack of four carbon steel boxes. The standard dynamic compaction was conducted on a 10-foot square grid pattern using both primary and secondary drops from a height of 42 feet of an 8-foot diameter, 20-ton weight to provide compaction within the center of each grid square (13,788 drop points). This resulted in the treatment of approximately 50% of the surface area. Standard dynamic compaction of this 1.5-acre trench resulted in the production of 5 to 6 foot craters with an average of 12 drops and final displacements between drops of less than 0.5 feet. (Phifer, 1991; Phifer and Serrato, 2000; Phifer and Wilhite 2001)

In 1998, standard dynamic compaction of 18-acres of the LLRWDF was conducted (primarily shallow land disposal units containing the typical stack of four carbon steel boxes). The standard dynamic compaction was conducted on a 10-foot square grid pattern using both primary and secondary drops from a height of 35 feet of an 8-foot diameter, 20-ton weight to provide compaction within the center of each grid square. Standard dynamic compaction at each drop location was conducted until the final displacement between drops of less than 0.25 feet was achieved, which may have required up to 35 drops per location. (Phifer and Serrato, 2000)

SRS Subsidence Treatment Conclusions and Resulting ELLWF Closure Implications

In 2001, SRNL conducted an evaluation of previous SRS static surcharge and dynamic compaction test programs and projects to evaluate the effectiveness of static surcharge and dynamic compaction on shallow land disposal units containing the typical stack of four carbon steel boxes (Phifer and Wilhite 2001). The following were the primary evaluation conclusions:

- The subsidence potential of the typical stack of four boxes containing low-density waste (i.e. 16 foot waste zone) is up to 13.5 feet.
- The subsidence potential is reduced by an estimated 10% for the static surcharge (25 feet) of shallow land disposal units containing the typical stack of four boxes containing low-density waste that are buried for less than 10 years. This means that a subsidence potential of up to 12.1 feet remains.
- The subsidence potential is reduced by an estimated 23% for standard dynamic compaction of shallow land disposal units containing the typical stack of four boxes containing low-density waste that are buried for less than 10 years. This means that a subsidence potential of up to 10.4 feet remains.
- Static surcharge and dynamic compaction of essentially intact boxes prior to significant corrosion has limited effectiveness due to the inherent structural stability of the matrix of boxes. Dynamic compaction of such boxes results in the lateral spread and physical interconnection of the top boxes, which may limit further dynamic compaction effectiveness.
- In order to increase the effectiveness of static surcharge and dynamic compaction on such stacked boxes, increased box corrosion and/or increased compactive effort is required. Increased compactive effort can be limited by the potential impact on surrounding structures.

Due to the subsidence potential, inherent structural stability of the matrix of boxes, and limited effectiveness of subsidence treatment prior to significant box corrosion, the closure plan for the ELLWF shallow land disposal units has been modified to include an interim closure phase. The interim closure phase consists of the installation of stormwater runoff covers, most likely an exposed geomembrane, which will be maintained until effective subsidence treatment is possible. At that point the subsidence treatment will be implemented followed by final closure. (Phifer et al. 2009) In order to determine the optimal subsidence treatment method and its timing, corrosion and subsidence treatment research and development is being conducted.

SRS Corrosion Studies

In 2001 a carbon steel box containing simulated low-density waste which was buried for 8 years was exhumed and evaluated for corrosion (see Figure 3). Based upon this one box corrosion evaluation, an estimation of box collapse without subsidence treatment was made. Box collapse due to corrosion is expected to occur between 50 and 500 years after burial and result in subsidence of the overlying materials. (Jones and Li 2001; Dunn 2001; Jones and Phifer 2002; Phifer et al. 2009) Three other simulated LLW boxes remain buried and will be exhumed and evaluated at future dates in order to narrow down the anticipated long-term corrosion rates and better estimate the optimal time of subsidence treatment.

An additional corrosion monitoring study was initiated in 2005 to further refine the time frame for effective subsidence treatment. Approximately 100 corrosion coupons made from a typical carbon steel box and a Sealand container were buried in shallow trenches at the ELLWF (see Figure 4). Two trenches of coupons are subjected to background infiltration and one trench is covered with high density polyethylene resulting in minimal infiltration. Subsequently the coupons will be exhumed and examined over time, in order to arrive at a long-term corrosion rate. Specially designed resistance measurement corrosion coupons are monitored on a quarterly basis to determine when coupons should be exhumed and examined.

Figure 3. Exhumed Carbon Steel Box

| a) Coupons in Trench | b) Covered Coupons |

Figure 4. Corrosion Coupon Monitoring Trench

Subsidence Treatment Modeling

Finite element modeling of the static loading of the typical stack of four carbon steel boxes at two time periods each for three different box corrosion scenarios has been conducted. Although the boxes are subjected to a static load, the structural response varies with time. The analytical results show that the deflection, buckling, and post buckling of the stacked boxes occur in sequence rather than simultaneously as corrosion progresses. Therefore, it is more appropriate to treat the problem as quasi-static rather than static. That is the structural response is dynamic with a very long duration. The modeling results for the short time frame corresponded well with the condition of the box exhumed and evaluated for corrosion as discussed above (see Figure 5). (Jones et al 2003; Wu et al 2005) This modeling exercise demonstrated the following:

- Static surcharge subsidence treatment of the stacked boxes can be effectively modeled
- The subsidence potential of the stacked boxes is a long-term cover system degradation mechanism
- Static surcharge subsidence treatment becomes more effective as box corrosion progresses

Finite element modeling was used to conduct a parametric study of the dynamic compaction of the typical stack of four carbon steel boxes. This study evaluated the mechanical behaviors associated with the strength of the boxes (i.e. corrosion), type of waste within the boxes (i.e. relative hardness), soil foundation and soil cover. In addition the study evaluated various kinematic boundary conditions, mesh sizes, and modeling and computational techniques to try and mimic known field results associated with dynamic compaction. (Gong 2001; Gong and Jones 2002) This modeling exercise demonstrated the following:

- Dynamic compaction subs dence treatment of the stacked boxes can be modeled (see Figure 5) but improved modeling tools and computer capacity should facilitate future modeling
- Dynamic compaction subsidence treatment becomes more effective as box corrosion progresses

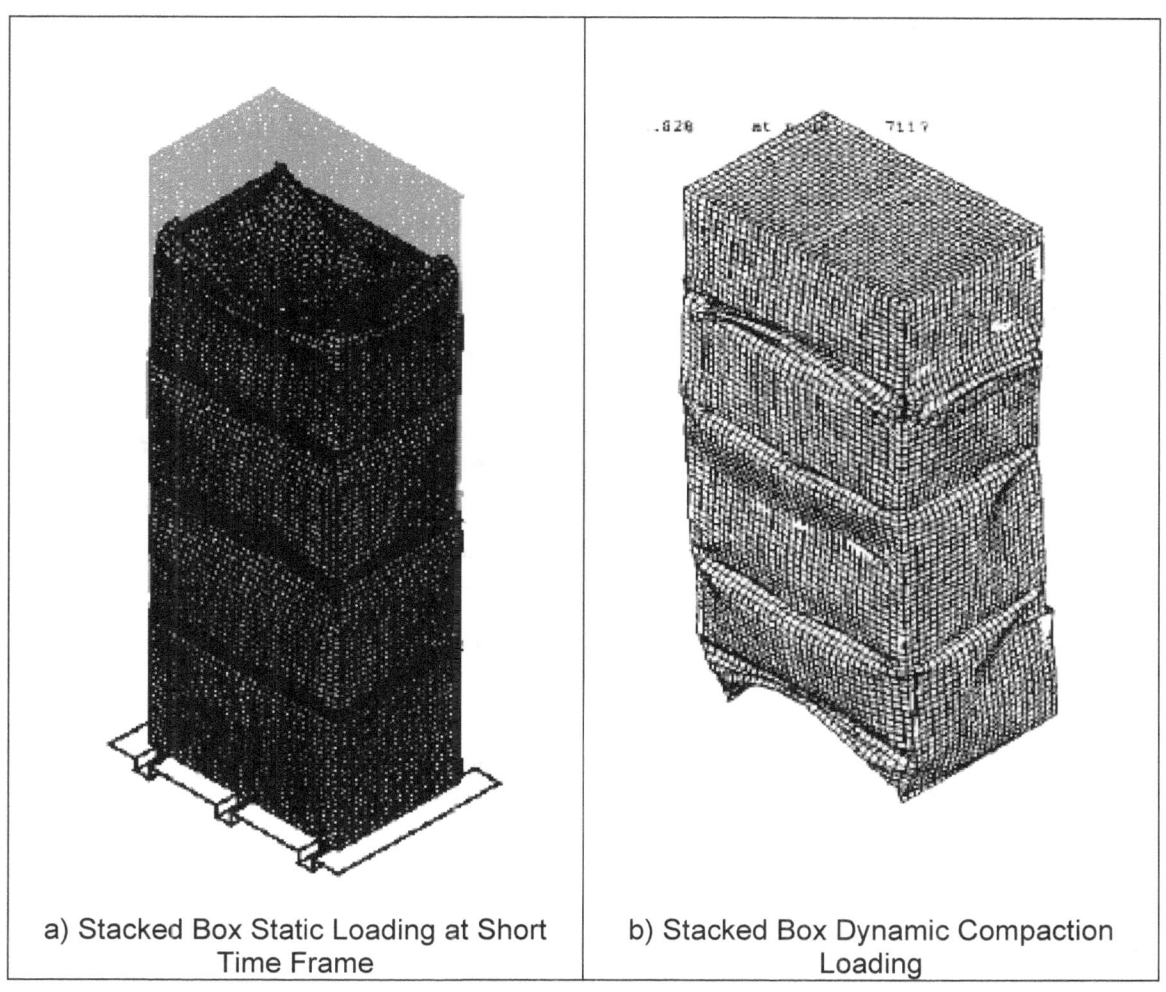

a) Stacked Box Static Loading at Short Time Frame	b) Stacked Box Dynamic Compaction Loading

Figure 5. Subsidence Treatment Modeling

Summary and Conclusions

Where applicable subsidence can be the primary degradation mechanism impacting the long-term performance of cover system engineered barriers. SRNL has conducted R&D related to subsidence in association with the ELLWF shallow land disposal units for over 20 years and has on-going subsidence R&D to determine the optimal subsidence treatment method and its timing. Additionally SRNL has contributed to the closure planning for the ELLWF shallow land disposal units to ensure that the subsidence potential is appropriately accommodated. Due to the subsidence potential, inherent structural stability of the matrix of boxes, and limited effectiveness of subsidence treatment prior to significant box corrosion, the closure plan for the ELLWF shallow land disposal units has been modified to include an interim closure phase. The interim closure phase consists of the installation of stormwater runoff covers, most likely an exposed geomembrane, which will be maintained until effective subsidence treatment is possible.

References

Dunn, K. A. 2002. B-25 Corrosion Evaluation Summary Report (U), WSRC-TR-2001-00587, Westinghouse Savannah River Company, Aiken, South Carolina. January 2002

Jones, W. E. and Li, W. T. 2001. Long-Term Waste Stabilization Parameter Estimation, WSRC-TR-2001-00323, Westinghouse Savannah River Company, Aiken, South Carolina. September 2001.

Jones, W. E. and Phifer, M. A. 2002. Corrosion and Potential Subsidence Scenarios for Buried B-25 Waste Containers (U), WSRC-TR-2002-00354. Westinghouse Savannah River Company, Aiken, South Carolina. September 2002.

Jones, W. E., Tsu-Te Wu, and M. A. Phifer. 2003. Structural Analysis for Subsidence of Stacked B-25 Boxes (U). Westinghouse Savannah River Company Technical Report, WSRC-TR-2002-00378, Savannah River Site, Aiken, SC.

Gong, C. 2001. Finite Element Analysis of Dynamic Compaction of a Stack of Four B-25 Boxes (U), WSRC-TR-2001-00320, Westinghouse Savannah River Company, Aiken, SC 29808. August 2001.

Gong, C. and Jones, W. E. 2002. Finite Element Analysis Enhancement for B-25 Container Dynamic Compaction (U), WSRC-TR-2002-00353, Westinghouse Savannah River Company, Aiken, SC 29808. September 2002.

McMullin, S. R. and Dendler, S. A. 1994. Dynamic Compaction Facility Test Report (U), WSRC-TR-94-0159, Westinghouse Savannah River Company, Aiken, SC 29808. March 1994.

Phifer, M. A. 1991. Closure of a Mixed Waste Landfill – Lessons Learned. Waste Management 91 Symposia, Tucson, Arizona pp. 517-525. 1991.

Phifer, M. A. and Serrato M. G. 2000. Preliminary E-Area Trench Subsidence Evaluation, SRT-EST-2000-00105. Westinghouse Savannah River Company, Aiken, South Carolina. January 18, 2000.

Phifer, M. A. and Wilhite, E. L. 2001. Waste Subsidence Potential versus Supercompaction, WSRC-RP-2001-00613. Westinghouse Savannah River Company, Aiken, South Carolina. September 27, 2001.

Phifer, M. A., Crapse, K. P., Millings, M. R., and Serrato M. G. 2009. Closure Plan for the E-Area Low-Level Waste Facility, SRNL-RP-2009-00075. Savannah River National Laboratory, Aiken, South Carolina. March 16, 2009.

Wu, T., Jones, W. E., and Phifer, M. A. 2005. Quasi-Static Analysis for Subsidence of Stacked B-25 Boxes, PVP2005-71537. Proceeding of 2005 ASME Pressure Vessels and Piping Division Conference, Denver, Colorado, July 17-21, 2005.

3.5.7   DOE Overview

Martin Letourneau

U.S. Department of Energy
Office of Environmental Management
Office of Environmental Compliance, EM-41
Martin.Letourneau@em.doe.gov

The Department of Energy (DOE) operates disposal facilities at several locations across the United States. Given that DOE disposal facilities are located on existing DOE sites, disposal concepts are developed using a systems based approach that considers the conditions at the site itself, engineered barriers, and waste forms as part of an integrated system to sufficiently isolate the waste. As a result, the disposal concepts regulated by DOE include a variety of different designs relying on different combinations of barriers.

DOE also operates disposal cells designed to accept mixed waste including large volume, uncontainerized wastes such as contaminated soils and rubble from remediation and decontamination and decommissioning activities. Although risk assessments and/or performance assessments (PAs) may be conducted for the facilities, disposal cells built as a result of a CERCLA Record of Decision are generally expected to meet specific regulatory design requirements (e.g., RCRA requirements). The contrasting design-based and performance-based viewpoints have resulted in the need for DOE to develop regulatory approaches that ensure that all of the applicable requirements of DOE, EPA and State regulations are met.

DOE Order 435.1 (USDOE 2001), Radioactive Waste Management, and the associated manual and guides are the directives for DOE-regulated radioactive waste disposal. A risk-informed, performance-based regulatory approach is used. The manual includes performance objectives for disposal facilities, a requirement for a performance assessment for each facility, and establishes the Low-Level Waste Disposal Facility Federal Review Group to conduct formal reviews of each PA after they have been prepared. The directives also include specific requirements for monitoring of disposal facilities, including recommendations related to comparing monitoring results with the results obtained in the PA. Requirements for PA maintenance are also included to provide a means to continue to build confidence in the conclusions of an assessment following approval to proceed with a disposal facility.

DOE Order 435.1 was originally issued in 1999 and a number of changes have occurred in the DOE system and lessons have been learned in the last 10+ years. In 2009, a process was initiated to update DOE Order 435.1 to address a number of areas of specific need (Letourneau et al. 2010). The first product of that process was a recently completed Complex-Wide Review that includes feedback from all of the DOE sites (USDOE 2010). Performance assessment is one area being considered for updates.

Two of the key areas for the update of interest for this workshop include: (1) an increased level of guidance regarding the role of probabilistic approaches for sensitivity and uncertainty analysis for PAs and (2) the role of "performance" monitoring, which is exploring ways to monitor a broader range of indicators of changes in performance in addition to the more traditional compliance monitoring, which emphasizes monitoring of concentrations of contaminants in water or air.

Experience has shown that PAs in support of waste disposal follow an iterative and learning process. Recognizing the need to be able to make decisions to operate a disposal facility in the presence of uncertainty, the concept of PA maintenance was formally identified as part of the development of DOE Order 435.1. PA maintenance includes any activities that are deemed necessary to confirm key assumptions and build confidence in the conclusions of a PA. This can include routine compliance monitoring, performance monitoring, laboratory or field experiments and demonstrations, and the conduct of supplemental analyses to address specific situations that may not have been covered by the PA (e.g., new or unique waste or waste form). Specific examples of PA maintenance activities from Savannah River (evaluation of cementitious materials, including the Cementitious Barriers Partnership), Idaho (laboratory column study), and Hanford (large scale cover demonstration) will be discussed in the presentation.

Monitoring activities are an important part of PA maintenance at all DOE disposal facilities. The manual and guides accompanying DOE Order 435.1 include specific recommendations to develop "action levels" for measurable parameters based on the results of the PA. In practice, this has resulted in establishing indicator concentrations at specific locations and times that are based on the results of the PA. These concentrations are then used as a point of comparison for actual monitoring results that are obtained each year. This approach is used to provide an indication that the facility is behaving within the acceptable envelope based on the PA. The concept of performance monitoring can provide additional points of comparison between the model results and the actual behavior of the disposal facility and also has the potential to allow for an indication of impending problems or confirmation of proper facility function sooner than more traditional monitoring methods. However, the goals of innovative performance monitoring approaches need to be properly described to the public. Performance monitoring tends to focus on learning more about the behavior of the system rather than being a specific comply/not comply decision point. This is an important distinction.

DOE has established a PA Community of Practice (CoP) (Letourneau et al. 2009, Seitz et al. 2010) to help improve the consistency of PA approaches being used and also more actively involve external stakeholders in the PA process. A key aspect of the CoP is to provide a means to share information before and during the process of conducting a PA to supplement the existing review role of the LFRG, which takes place after the PA has been conducted. The PA CoP includes support for activities such as workshops on specific topics of interest, PA assistance teams that provide a means for expertise from one site to be transferred to another as a PA is being conducted, support for scoping processes that allow regulators and other stakeholders to have input to the development of the PA, and also encouraging the sharing of information. The LFRG has encouraged broader sharing of information related to PAs and expects that DOE PAs will be available to the public. Tank closure-related PAs are available on the DOE EM website and a number of site-specific PAs and supporting information are available on the DOE Information Bridge.

References

USDOE, Complex-Wide Review of DOE's Radioactive Waste Management, US Department of Energy, Office of Environmental Management, Washington, DC (2010).

Letourneau, M.J., et al., The United States Department of Energy Radioactive Waste management Order 435.1 Update, The First Step – A Complex Wide Review, Proceedings from WM 2010, March 7-11, Phoenix, AZ (2010).

Letourneau, M.J., et al., Improving Consistency of Performance Assessments in the USDOE Complex, Proceedings from WM 2009, March 1-5, Phoenix, AZ (2009).

Seitz, R.R. et al., Performance Assessment Community of Practice, Proceedings from WM 2010, March 7-11, Phoenix, AZ (2010).

USDOE, Radioactive Waste Management, DOE O 435.1 – Chg 1, US Department of Energy, Washington DC (2001).

SESSION 2:

DEGRADATION PROCESSES AND PERFORMANCE EVOLUTION OF ENGINEERED BARRIERS

Session Chairs:
Craig Benson, University of Wisconsin/CRESP and
W. Jody Waugh, S.M. Stoller LLC

Technical Reporter:
Brooke Traynham, NRC/FSME

4.1   Motivation

Physical, biological, and chemical processes can induce changes in the structure, physical, and biological characteristics of covers that are intrinsic to their proper functioning as barrier systems. Degradation processes can affect barrier components (e.g., compacted soil layers, geomembranes, GCLs, drainage layers) and the overall performance in many ways. The goals of this session were to discuss how our understanding of degradation processes may be used to improve the design and performance of covers and liners. This session presents an overview of the most significant short-term and long-term degradation processes causing increases in radon release, water percolation, erosion, and bio-uptake, as well as, potential mechanisms to prevent or mitigate cover degradation.

4.2   Background

Nature alters engineered soil covers in short-time periods; for example, hydraulic properties of engineered fine-textured soil layers become more similar over time to the surrounding environmental system, regardless of the initial conditions. Degradation processes occur over a broad range of time scales and include, but are not limited to, climatic variability, plant succession, geomorphic processes, pedogenesis, anthropogenic impacts, erosion, microbial processes that affect barrier materials and drains (e.g., biofouling), and geochemical processes. Strategies to reduce one degradation process may cause unintended impacts and enhance other degradation processes; for example, activities that reduce erosion inadvertently may cause an increase in water percolation. Similarly, the current practice of using herbicides to control vegetation may actually be detrimental by inhibiting shrub establishment. The shrubs increase transpiration, and promote soil development to increase water storage. Hence, in some cases, ecological succession (e.g., shrub establishment) and soil development may be a solution rather than a problem for water balance covers.

4.3   Challenges in Prediction of Cover Performance

Degradation processes include the following physical, chemical, and biological processes:

(1) Physical processes such as freeze-thaw, wet-dry cycle, differential settlement, retention of burrow soil structure (peds or clods) during construction, UV degradation, thermal degradation, erosion, fire, and pedogenesis.

(2) Chemical processes such as oxidation of geosynthetic materials and cation exchange mechanisms in sodium bentonites.

(3) Biological processes that involve unanticipated ecological consequences of designs that, by creating habitat for deep-rooted plants, burrowing animals, and soil microorganisms, can alter soil hydraulic properties.

The challenge of predicting cover performance in the context of the aforementioned degradation processes is confounded by: (a) the need to infer long-term conditions from short-term monitoring data; (b) difficulty in scaling from bench to field; (c) data gaps; (d) developing and screening future scenarios;, (e) standardize future intruder scenarios; (f) unforeseen ecological consequences; and (g) the tendency of root intrusion and soil development to alter hydraulic properties.

Understanding the range of time and spatial scales over which different degradation processes operate is important when characterizing degradation processes. Timescales of 100s to 1000s of years need to be considered. Spatial heterogeneity of the system also needs consideration.

4.4   Recommendations

Strategies for minimizing the negative impacts of degradation processes include; (a) careful attention to construction quality assurance (QA); (QA is especially important to the successful short-term performance of the cover); (b) identification of the processes that have the greatest impact on performance; (c) understanding the total system and planning accordingly; (d) analysis of each component within context; and (e) identification and standardization of scenarios that ultimately control the performance objectives and cover features related to these objectives.

Unintended consequences must be considered when trying to diminish one form of degradation. The function of each cover component (e.g., use of plants to inhibit deep percolation) must be clearly determined to diminish the risk of unintended consequences. The intended function of each component must be clearly defined in addition to the potential risks to the primary function. This will assist in delineating pathways for managing the total system performance.

Designers and analysts should recognize that soil properties may change quickly, and therefore should minimize the consequences of these changes by designing and constructing covers that mimic longer-term conditions that are congruent with nature. The resources required to maintain the engineered system must also be considered. Instruments can help identify the timing of a degradation process. There is a need to develop techniques that seek to understand magnitude and direction of natural changes anticipated to occur. One approach is to develop a catalog of natural analogs.

Engineered systems evolve towards a natural equilibrium. Thus, over time, cover properties tend to become more similar across sites no matter how differently the systems were initially engineered and built. Cover degradation attributable to pedogenesis and ecological change should be recognized as an inevitable, fairly predictable, natural succession. The term "degradation" implies a loss of performance. In some cases, however, natural pedogenesis and ecological succession can lead to improved system performance over time. Therefore, performance will be more steady and characteristic over time, if engineered soil layers and vegetation are designed to more closely resemble characteristics of natural systems. Similarly, cover renovation designs should recognize this behavior, and should be selected so that long-term maintenance needs are minimized. Human and environmental disturbances have a greater impact on systems that are not in natural equilibrium, where natural feedbacks have been disabled. Cover design and monitoring plans must be guided with this endpoint in mind.

Future research should focus on: (1) using natural analogs to better understand and evaluate long-term degradation processes including both spatial heterogeneity and temporal trajectories of change; (2) designing covers that mimic the favorable attributes of selected natural analogs; (3) evaluating effects of soil development and ecological change on radon exhalation and bio-uptake pathways; (4) evaluating effects of waste subsidence on long-term cover performance. and (5) predicting and incorporating landform changes in cover and disposal cell designs.

4.5   Extended Abstracts

One of the more significant developments with regards to engineered barriers near the surface is the recognition by engineers and scientists alike of the speed in which certain components of the barrier can change their original as-built properties. This recognition has led to increased interest and research into all aspects of the diverse processes that cause degradation of engineered properties. Leading experts in this field accepted invitations by the organizing committee to present and submitted extended abstracts on the developments and status of this research.

4.5.1 DOE Experience with Cover Degradation Processes, Design Improvements, and Cover Renovation for Uranium Mill Tailings Disposal Cells

W. J. Waugh

S.M. Stoller Corporation

Introduction

The U.S. Department of Energy (DOE) is responsible for long-term surveillance and maintenance (LTSM) of disposal cells constructed to contain tailings and other residual milling materials as mandated primarily through the Uranium Mill Tailings Radiation Control Act of 1978 (UMTRCA). The key feature of UMTRCA disposal cells is an engineered cover designed to imit release of radon into the atmosphere, and at some sites, to limit percolation of rainwater and leaching of contaminants into groundwater—and to continue to do so in perpetuity. DOE routinely conducts visual inspections of the disposal cells as part of LTSM and intermittently conducts follow-up investigations of cover degradation processes and performance as a best management practice.

In response to the workshop questionnaire, this presentation briefly reviews (1) the regulatory framework for LTSM of UMTRCA covers, (2) follow-up investigations of cover degradation processes and their effects on cover performance, (3) resulting improvements in the design and performance of covers, and (4) an ongoing study of cover renovation as an option for reducing long-term effects of degradation processes on cover performance.

Regulatory Framework and Standards

UMTRCA authorized the U.S. Environmental Protection Agency (EPA) to issue standards for remedial action, groundwater quality, and the performance of disposal cells. The act designated the U.S. Nuclear Regulatory Commission (NRC) as the agency responsible for enforcing EPA standards and assigned the responsibility for remediation and long-term care to DOE.

In 1983, EPA issued standards for disposal cell longevity and radon attenuation (EPA 1983). The longevity target was set at 1,000 years with a minimum performance period of 200 years. The radon standard requires reasonable assurance that releases of radon-222 to the atmosphere will not exceed an average surface flux rate of 20 pCi m^{-2} s^{-1}. In 1985, the 10th Circuit Court of Appeals set aside the groundwater provisions of UMTRCA and remanded them to the EPA. In 1987, EPA published draft groundwater standards, and in 1989, DOE began incorporating the draft standards under the assumption that they would soon become final. In 1995, EPA published standards or concentration limits for amounts of radioactive and associated hazardous constituents in groundwater (EPA 1995).

UMTRCA requires NRC (1) to evaluate designs and implementation of remedies whether by DOE (under UMTRCA Title I) or by private site owners (under UMTRCA Title II) and (2) after remediation, to concur that remedies satisfy the standards mandated by EPA. Once NRC makes the determination that applicable EPA standards and requirements for remedial action, disposal cell performance, and groundwater quality have been satisfied, DOE becomes a licensee to NRC for LTSM site inspections, monitoring, reporting, and record-keeping.

Evolution of Cover Designs

Cover design concepts, construction practices, and performance evaluations have undergone significant changes since passage of UMTRCA. Design improvements primarily reflected (1) efforts to satisfy EPA standards for groundwater quality, (2) a better understanding of degradation processes, and (3) overall advances in the state of the science (Waugh et al. 2001).

Before EPA mandated groundwater quality standards, the cover design process focused on radon attenuation and longevity. Early designs consisted basically of three layers: (1) a compacted soil layer (CSL) or radon barrier overlying the tailings (NRC 1989), (2) a rock riprap layer at the surface to control erosion (NRC 1990), and (3) a bedding layer of coarse sand or small gravel sandwiched between the CSL and riprap. Disposal cell covers at Burrell, Pennsylvania; Lakeview, Oregon; Shiprock, New Mexico; and Tuba City, Arizona, are examples of the early designs (http://www.lm.doe.gov).

In anticipation of EPA groundwater quality standards, DOE estimated, retrospectively, that CSLs in the early covers (1) had a saturated hydraulic conductivity of approximately 10^{-7} cm s^{-1}, (2) would operate at an unsaturated hydraulic conductivity of 10^{-9} cm s^{-1}, and (3) that "any moisture that penetrated the upper portion of the radon barrier is held there until it is evaporated"; thus, "this design prevents the radon barrier from becoming saturated" (DOE 1988). DOE also developed a new design framework that placed greater emphasis on limiting percolation flux in the CSL (DOE 1989). The framework recommended a conservative percolation flux for CSLs that is equivalent to a saturated hydraulic conductivity between 10^{-7} and 10^{-8} cm s^{-1}. The framework included an option for placing a soil layer above the CSL to protect it from degradation processes such as freeze/thaw cracking and biointrusion that could increase hydraulic conductivity. Colorado examples of this improved, low-permeability design are the Durango, Gunnison, Grand Junction, and Rifle covers (http://www.lm.doe.gov).

Advances in the science of cover designs and lessons learned from observing and evaluating degradation processes in both the early covers and the later, low-permeability covers (see next section) contributed to DOE's development of alternative designs (Waugh et al. 2001). In many arid and semiarid ecosystems, relatively low precipitation, high potential evapotranspiration (ET), and thick unsaturated soils limit deep percolation (Gee and Tyler 1994). DOE and others developed covers that mimic this natural water conservation, sometimes called ET covers, in an attempt to provide more sustainable tailings isolation and groundwater protection (Nyhan et al. 1990; Anderson et al. 1993; Ward and Gee 1997; Waugh and Richardson 1997; Albright et al. 2004; Scanlon et al. 2005). The disposal cell at Monticello, Utah (http://www.lm.doe.gov), a uranium mill tailings site regulated under RCRA/CERCLA, is an example of this alternative design. To limit percolation, the Monticello cover relies on a 160-cm layer of sandy clay loam soil overlying a 40-cm sand capillary barrier as a water storage layer, and a planting of native sagebrush steppe vegetation to seasonally remove soil water and limit percolation (Waugh et al. 2009).

Cover Degradation Processes and Performance

This section is a brief summary of pertinent literature on degradation processes in covers, DOE observations and evaluations of interrelated ecological change and soil development processes, and their effects on the performance of uranium mill tailings covers. Several studies have shown that CSLs in conventional covers often fall short of low-permeability targets, often during or shortly after construction, and sometimes by several orders of magnitude. Several reasons related to construction and degradation processes are cited:

- Differences between laboratory and field saturated hydraulic conductivities, and compaction of CSLs either dry or wet of optimum during construction (Benson et al. 1999; Daniel 1984).
- Freeze-thaw and desiccation cracking (Benson et al. 1999; Kim and Daniel 1992; Benson and Othman 1993).
- Differential settlement (Jessberger and Stone 1991; Lagata 1992).
- Unanticipated ecological consequences of designs that encourage biointrusion and soil development (Hakonson 1986 Suter et al. 1993; Bowerman and Redente 1998; Waugh et al. 1999).
- Retention of borrow soil structure (peds or clods) during construction and pedogenesis (soil development processes) after construction (Albright et al., 2006; Waugh et al., 2007; Benson et al., 2010).

Deep-rooted plants began growing on UMTRCA rock covers within a few years after construction (DOE 1992). Surface layers of rock or gravel reduce evaporation (Groenevelt et al. 1989), increase soil water storage (Kemper et al. 1994; Sackshewsky et al. 1995), and consequently create habitat for deep-rooted plants. Accumulation of windblown dust in the West and leaf litter in the East has accelerated soil development and ecological succession n rock riprap layers. Plant roots may concentrate in and extract water from compacted clay layers, causing desiccation cracking, even when overlying soils are nearly saturated (Hakonson 1986), indicating that the rate of plant water extraction may exceed the rehydration rate of the compacted clay. Roots can also clog lateral drainage layers (DOE 1992). Plants rooted in uranium mill tailings may contain elevated levels of U, Mo, Se, ^{226}Ra, ^{230}Th, and ^{210}Po (Dreesen and Williams 1982; Hosner et al. 1992; Markose et al. 1993). Radon-222 can be transported into the atmosphere as plant roots extract water from tailings (Lewis and MacDonell 1990; Morris and Fraley 1989). Roots may also alter waste chemistry, potentially mobilizing contaminants (Cataldo et al. 1987).

DOE tracked woody plant roots down into or through CSLs at the Burrell, Durango, Grand Junction, Lakeview, Shiprock, and Tuba City disposal cells (DOE 1992; Waugh et al. 1999, 2007). Tap roots typically extended vertically through the rock riprap and bedding layers and then branched and spread laterally at the CSL surface. Secondary and tertiary roots extended vertically into the CSL, where they became fibrous root mats following cracks and soil structural planes.

In a follow-up investigation of root intrusion, DOE evaluated effects of plant roots and soil development on in situ saturated hydraulic conductivity (K_{sat}) of CSLs at Burrell, Lakeview, Shiprock, and Tuba City using air-entry permeameters (Stephens et al. 1988). At Burrell, the mean K_{sat} (n=3) was 3.0×10^{-5} cm s^{-1} where Japanese knotweed roots penetrated the CSL, compared to 2.9×10^{-7} cm s^{-1} at locations with no plants (Waugh et al. 1999). The weighted-average K_{sat} for the entire cover, calculated using the community leaf area index for Japanese knotweed (Wells and Norman 1991), was 4.4×10^{-6} cm s^{-1}. At Lakeview, the mean K_{sat} (n=17) for the CSL both with and without sagebrush and bitterbrush roots, was 3.0×10^{-5} cm s^{-1} (Waugh et al. 2007). The highest K_{sat} values occurred near the top of the CSL; the lowest values occurred deeper in the CSL. The mean K_{sat} (n=6) in the top of the Shiprock CSL was 4.4×10^{-5} cm s^{-1} (Glenn and Waugh 2001). Results were highly variable and lower where tamarisk and Russian thistle were rooted in the CSL. The Shiprock CSL was nearly saturated, as measured monthly for 16 months at four locations using a neutron hydroprobe (Gardner 1986). At Tuba City, the mean K_{sat} (n=16) of the CSL was 8.7×10^{-6} cm s^{-1}, and values ranged from a low of 9.8×10^{-9} to a high of 1.18×10^{-4} cm s^{-1}. In all the above tests, dyes indicated that water moved through macropore cracks in the soil structure of CSLs.

DOE also investigated effects of soil development and changes in plant ecology on the performance of the ET cover at Monticello (Waugh et al. 2009). Two effects were of interest: (1) whether development of soil structure can cause percolating water to pass through and essentially reduce the storage capacity of the water storage layer, and (2) whether soil development would loosen the storage layer and result in higher porosity and greater water storage capacity.

An evaluation of soil morphology, soil hydraulic properties, and plant ecology of the Monticello cover 7 years after construction reached the following conclusions (Benson et al. 2008):

- The water storage layer of the cover had developed weak-to-moderate prismatic and platy soil structure with some vertical factures, although structural development varied greatly with depth and from one location to another.
- Differences in soil structure from one location to another had little influence on the hydraulic conductivity of the water storage layer.
- Saturated volumetric water content of the water storage layer had increased, and the air-entry suction had decreased over the 7 years; however, differences in soil structure from one location to another had little influence on soil water retention characteristics.
- Differences in the abundance and growth of shrubs from one location to another supported the hypothesis that greater soil structural development at depth contributes to a greater abundance of deeper-rooted shrubs, including sagebrush and rabbitbrush. These shrubs may be important for water extraction lower in the water storage layer of the Monticello cover.
- Overall, this greater-than-expected soil structural development 7 years after construction had little influence on the overall performance of the cover. The cumulative percolation in the Monticello ET cover, monitored over a 9-year period in a 3-ha embedded lysimeter, was 4.8 mm, with over 80 percent of the total occurring during one very wet year (Waugh et al. 2009).

Cover Renovation

In 2008, DOE constructed a research facility near Grand Junction, Colorado, to test low-cost methods for renovating conventional covers and to compare the performance of conventional and renovated covers. Without intervention, soil development and ecological succession will, over time, effectively transform existing conventional covers into ET covers. Renovation offers the potential to accelerate and enhance this transformation. Hence, by accommodating degradation processes, the goal is to sustain a high level of performance and reduce LTSM costs over the long term. LTSM often includes controlling vegetation on conventional rock riprap covers. The cost of vegetation management has increased as a more favorable habitat has developed for deep-rooted shrubs.

Renovation may involve deliberately ripping and blending the rock riprap, bedding layer, and underlying CSL and planting native shrubs in the rip rows. Geomorphological and ecological evidence from a nearby analog site suggests that the resulting vegetated rocky slope should be stable for hundreds if not thousands of years and have a favorable water balance (Smith et al. 1997).

The cover renovation test facility consists of two large drainage lysimeters constructed using the methods of Albright et al. (2004). One of the test sections will be renovated, whereas the other will be monitored as a control.

The cover renovation investigation will include contaminant release processes other than percolation. Effects of cover degradation processes and of cover renovation methods on radon attenuation, biological intrusion, contaminant uptake, erosion, and slope stability will also be evaluated.

References

Albright, W.H., C.H. Benson, G.W. Gee, A.C. Roesler, T. Abichou, P. Apiwantragoon, B.F. Lyles, and S.A. Rock, 2004. "Field water balance of landfill final covers," *Journal of Environmental Quality*, 33: 2317–2332.

Albright, W., C. Benson, G Gee, T. Abichou, S. Tyler, and S. Rock, 2006. "Field performance of three compacted clay landfill covers," *Vadose Zone J.*, 5:1157–1171.

Anderson, J.E., S. Nowak, T.D. Ratzlaff, and O.D. Markham, 1993. "Managing soil moisture on waste burial sites in arid regions," *Journal of Environmental Quality*, 22: 62–69.

Benson, C.H., and M.A. Othman, 1993. "Hydraulic conductivity of compacted clay frozen and thawed in situ," *Journal of Geotechnical Engineering*, 119:276–294.

Benson, C.H., D.E. Daniel, and G.P. Boutwell, 1999. "Field performance of compacted clay liners," *Journal of Geotechnical and Geoenvironmental Engineering,* 125: 390.

Benson, C.H., S.H. Lee, X. Wang, W.H. Albright, and W.J. Waugh, 2008. *Hydraulic properties and geomorphology of the earthen component of the final cover at the Monticello uranium mill tailings repository,* Geo Engineering Report No. 08-04, Geological Engineering, University of Wisconsin, Madison, Wisconsin.

Benson, C.H., W.H. Albright, D.O. Fratta, J.M Tinjum, E. Kucukkirca, S.H. Lee, J. Scalia, P.D. Schlicht, and X. Wang, 2010 (DRAFT). *Engineered Covers for Waste Containment: Changes in Engineering Properties and Implications for Long-Term Performance Assessment*, NUREG/CR-XXXX, Office of Research, U S. Nuclear Regulatory Commission, Washington, DC.

Bowerman, A.G., and E.F. Redente, 1998. "Biointrusion of protective barriers at hazardous waste sites," *Journal of Environmental Quality*, 27 625–632.

Cataldo, D.A., C.E. Cowan, K.M. McFadden, T.R. Garland, and R.E. Wildung, 1987. *Plant rhizoshpere processes influencing radionuclide mobility in soil,* PNL–6277, Pacific Northwest National Laboratory, Richland, Washington.

Daniel, D., 1984. "Predicting hydraulic conductivity of clay liners," *Journal of Geotechnical Engineering*, ASCE, 110: 285–300.

DOE, 1988. *Moisture Content and Unsaturated Conditions in UMTRA Project Radon Barriers,* DOE/UMTRA-400656, United States Department of Energy, Albuquerque, NM.

DOE, 1989. *Technical Approach Document, Revision II,* UMTRA-DOE/AL 050425.0002, United States Department of Energy, Albuquerque NM.

DOE, 1992. *Vegetation growth patterns on six rock-covered UMTRA Project disposal cells*, DOE/UMTRA 400677-0000, U.S. Department of Energy, Albuquerque, NM.

Dreesen, D.R., and J.M. Williams, 1982. "Mobility and bioavailability of uranium mill tailings contaminants," *Environ. Sci. Technol.*, 16: 702–709.

EPA (U.S. Environmental Protection Agency), 1983. "Health and Environmental Protection Standards for Uranium and Thorium Mill Tailings," Title 40 *Code of Federal Regulations* Part 192.

EPA (U.S. Environmental Protection Agency), 1995. "Groundwater Standards for Remedial Actions at Inactive Uranium Processing Sites; Final Rule," *Federal Register* Vol. 60 No. 7, p.2854, Wednesday January 11, 1995 (40 *Code of Federal Regulations* Part 192).

Gardner, W.H., 1986. "Water content," pp. 493–544, in A. Klute (ed.), *Methods of soil analysis, part I, physical and mineralogical methods,* second edition, American Society of Agronomy and Soil Science Society of America, Madison, Wisconsin.

Gee, G.W., and S.W. Tyler (eds.), 1994. "Symposium: Recharge in arid and semiarid regions," *Soil Science Society of America Journal,* 58: 5–72.

Glenn, E., and J. Waugh, 2001. *Disposal Cell Cover Moisture Content and Hydraulic Conductivity, Long-term Surveillance and Maintenance Program, Shiprock, New Mexico, Site,* GJO-2001-204-TAR, U.S. Department of Energy, Grand Junction, Colorado.

Groenevelt, P.H., P. van Straaten, V. Rasiah, and J. Simpson, 1989. "Modification in evaporation parameters by rock mulches," *Soil Technol.,* 2: 279–285.

Hakonson, T.E., 1986. *Evaluation of geologic materials to limit biological intrusion into low-level radioactive waste disposal sites,* LA-10286-MS, Los Alamos National Laboratory, Los Alamos, New Mexico.

Hosner, L.R., H.J. Woodard, and J. Bush, 1992. "Growth and selenium uptake of range plants propagated in uranium mine soils," *J. Plant Nutrition,* 15: 2743–2761.

Jessberger, H.L., and K. Stone, 1991. "Subsidence effects on clay barriers," *Geotech,* 41: 185–194.

Kemper, W.D., A.D. Nicks, and A.T. Corey, 1994. "Accumulation of water in soils under gravel and sand mulches," *Soil Sci. Soc. Am. J.,* 58: 56–63.

Kim, W.H., and D.E. Daniel, 1992. "Effects of freezing on the hydraulic conductivity of compacted clay," *Journal of Geotechnical Engineering,* 18: 1083–1097.

Lagata, M.D., 1992. *Hydraulic conductivity tests on geosynthetic clay liners subjected to differential settlement,* MS Thesis, Univ. of Texas, Austin, Texas.

Lewis, B.G., and M.M. MacDonell, 1990. "Release of radon-222 by vascular plants: Effect of transpiration and leaf area," *J. Env. Radioactivity,* 19: 93–97.

Markose, P.M., I.S. Bhat, and K.C. Pillai, 1993. "Some characteristics of ^{226}Ra transfer from soil and uranium mill tailings to plants," *J. Environ. Radioactivity,* 21:131–142.

Morris, R.C., and L. Fraley, Jr, 1989. "Effects of vegetation, a clay cap, and environmental variables on Rn-222 fluence rate from reclaimed U mill tailings," *Health Physics,* 56: 431–440.

NRC, 1989. *Calculation of Radon Flux Attenuation by Earthen Uranium Mill Tailings Covers,* Regulatory Guide 3.64 (Task WM 503–4), U.S. Nuclear Regulatory Commission, Washington, DC.

NRC, 1990. *Staff Technical Position: Design of Erosion Protection Covers for Stabilization of Uranium Mill Tailings Sites,* NRC/UMT/10030064, U.S. Nuclear regulatory Commission, Washington, DC.

Nyhan, J.W., T.E. Hakonson, and B J. Drennon, 1990. "A water balance study of two landfill cover designs for semiarid regions," *Journal of Environmental Quality,* 19: 281–288.

Sackshewsky, M.R., C.J. Kemp, S.O. Link, and W.J. Waugh, 1995. "Soil water balance changes in engineered soil surfaces," *Journal of Environmental Quality* 24: 352–359.

Scanlon, B.R., R.C. Reedy, K.E. Keese, and S.F. Dwyer, 2005. "Evaluation of evapotranspirative covers for waste containment in arid and semiarid regions in the southwestern USA," *Vadose Zone Journal,* 4: 55–71.

Smith, G.M., W.J. Waugh, and M.K. Kastens, 1997. "Analog of the long-term performance of vegetated rocky slopes for landfill covers," pp. 291-300, in *Tailings and Mine Waste 1997,* A.A. Balkema, Rotterdam.

Stephens, D.B., M. Unruh, J. Havlena, R.G. Knowlton, Jr., E. Mattson, and W. Cox, 1988. "Vadose zone characterization of low-permeability sediments using field permeameters," *Ground Water Monitoring Rev.,* 8: 59–66.

Suter II, G.W., R.J. Luxmoore, and E.D. Smith 1993. "Compacted soil barriers at abandoned landfill sites are likely to fail in the long term," *Journal of Environmental Quality*, 22: 217–226.

Ward, A.L., and G.W. Gee, 1997. "Performance evaluation of a field-scale surface barrier," *Journal of Environmental Quality*, 26: 694–705.

Waugh, W.J., and G.N. Richardson, 1997. "Ecology, design, and long-term performance of surface barriers: Applications at a uranium mill tailings site," pp. 36–49, in *Barrier Technologies for Environmental Management,* National Research Council, National Academy Press.

Waugh, W.J., S.J. Morrison, G.M. Smith, M. Kautsky, T.R. Bartlett, C.E. Carpenter, and C.A. Jones, 1999. *Plant Encroachment on the Burrell, Pennsylvania, Disposal Cell: Evaluation of Long-Term Performance and Risk,* GJO-99-96-TAR, U.S. Department of Energy, Grand Junction, Colorado.

Waugh, W.J., G.M. Smith, D.Bergman-Tabbert, and D.R. Metzler, 2001. "Evolution of cover systems for the Uranium Mill Tailings Remedial Action Project, USA," *Mine Water and the Environment*, 20: 190–197.

Waugh, W.J., G.M. Smith, B. Danforth, G.W. Gee, V. Kothari, and T. Fauling, 2007. "Performance evaluation of the engineered cover at the Lakeview, Oregon, uranium mill tailings site," Proceedings of Waste Management 2007 Symposium, Tucson, Arizona.

Waugh, W.J., C.H. Benson, and W.H. Albright, 2009. "Sustainable covers for uranium mill tailings, USA Alternative design, performance, and renovation," Proceedings of 12th International Conference on Environmental Remediation and Radioactive Waste Management, Liverpool, UK.

Wells, J.M., and J.M. Norman, 1991. "Instrument for indirect measurement of canopy architecture," *Agron. J.*, 83: 818–825.

4.5.2 Processes that Alter the Structure of Soils and Their Effects on the Performance of Covers

Craig H. Benson, PhD, PE, DGE

Wisconsin Distinguished Professor of Geological Engineering
University of Wisconsin-Madison
www.gle.wisc.edu
chbenson@wisc.edu

The final cover controls the amount of liquid entering a waste containment system and therefore directly affects the amount of leachate generated within the waste mass and the potential flux of contaminants to ground water. As a result, the final cover generally is regarded as one of the most important components in a waste containment system with respect to long-term performance. This is particularly true for radioactive wastes, which can pose a very long-term threat to ground water. Consequently, considerable interest has developed in understanding the long-term performance of final covers and the processes that affect the engineering behavior of earthen and geosynthetic layers used in cover systems (Albright et al. 2010).

Physical, biological, and chemical processes can induce changes in the structure and physical characteristics of soils that are intrinsic to their proper functioning in barrier systems (Suter et al. 1993, Benson et al. 2007). Wet-dry cycling and freeze-thaw cycling are the primary physical processes that cause changes in structure and alteration of physical properties. Both mechanisms alter the water pressure in soils, which induces shrinkage, tensile strains, and cracks where the tensile strength of the soil has been exceeded (Chamberlain and Gow 1979, Benson and Othman 1993; Benson et al. 1995; Albrecht and Benson 2001, 2002). Ice that forms during freezing also causes cracking and secondary structure. Cracks may form on preexisting planes of weakness or in material carefully remolded during soil placement. The cracks create secondary structure in soil, which alters the soil hydraulic properties and the performance of a final cover. Wet-dry and freeze-thaw cycling are generally considered to be physical processes. However, wet-dry cycling can be induced by through root water uptake, which is a biologically driven process.

Chemical processes may also affect the structure and hydraulic properties of soils used in covers. The most significant are cation exchange mechanisms in sodium bentonites used in hydraulic barrier layers. These exchange processes alter the swelling capacity and hydraulic properties of bentonites appreciably (Shackelford et al. 2000, Jo et al. 2001). In some cases, a very impermeable bentonitic barrier layer can be transformed to a highly permeable layer (Egloffstein 2001, Benson et al. 2007, Meer and Benson 2007, Scalia and Benson 2010). Precipitation and oxidation processes may also be important through the deposition of solids on surfaces exposed to water and/or gas flow (Eigenbrod 2003). These solids may plug pores, resulting in alterations in hydraulic properties. However, these processes have not been documented to a significant degree in final covers.

The effect of these processes, and the resulting alterations in soil structure, on the performance of final covers has been recognized. Although numerous studies to evaluate these soil-altering mechanisms have been conducted in the laboratory and field, no studies had been conducted until recently where alterations in earthen materials and the performance of final covers were evaluated in an integrated study (Albright et al. 2006). Consequently, considerable uncertainty has existed in how to account for alterations in soil structure and soil hydraulic properties in modeling conducted for performance assessments.

During the last decade, several studies were initiated in the US and abroad to assess the long-term performance of modern final covers under realistic conditions (Melchior 1997, Dwyer 2001, Albright et al. 2004). The most comprehensive of these studies was the Alternative Cover Assessment Program (ACAP), which evaluated the performance of 27 different final covers at 12 locations in 8 states in the US (Albright et al. 2004). The ACAP covers were evaluated for a period ranging from 4-8 yr in climates ranging from arid (Apple Valley, CA) to humid (Albany, GA) to seasonal semi-arid with extreme fluctuations in temperature (Underwood, ND). Conventional covers relying on hydraulic barriers (compacted clay barriers, geosynthetic clay liners, polymeric geomembranes, and combinations thereof) and alternative covers relying on water balance principles were evaluated in ACAP via extensive monitoring of the water balance, state variables, and meteorological conditions.

At the end of the study, the ACAP test facilities were exhumed to evaluate how the properties of the cover materials changed 4.0-8.9 yr after installation (6.3 yr on average). Field tests were conducted, samples were collected, laboratory testing was performed, and data analyses were conducted to evaluate how the properties of earthen cover materials changed while in service Thus, ACAP also provided a systematic coupling of cover performance and alterations in cover soils and their hydraulic properties (e.g., Albright et al. 2006).

Formation of soil structure changed the hydraulic properties in nearly all cover soils evaluated by ACAP, regardless of climate, cover design, or service life (Benson et al. 2010). The saturated hydraulic conductivity and the ☐ parameter for the soil water characteristic curve (SWCC) increased, which reflects formation of larger pores due to pedogenic processes such as wet-dry and freeze-thaw cycling. Larger changes were observed for soils with lower as-built saturated hydraulic conductivity and soils with a greater proportion of clay particles in the fines fraction. Hydraulic properties of the cover soils were similar when exhumed, regardless of the as-built condition (Benson et al. 2007, 2010). These changes are consistent with those observed in past studies evaluating wet-dry and freeze-thaw mechanisms (Benson and Othman 1993, Albrecht and Benson 2001 2002). Albright et al. (2006) illustrate how these changes affect the performance of final covers.

Some of the geosynthetic clay liners (GCLs) exhibited significant changes in chemical and physical properties similar to those reported in previous mechanistic studies. GCLs were found to retain very low saturated hydraulic conductivity (< 5×10^{-11} m/s) when placed on a moist subgrade (water content > optimum water content) and covered with a geomembrane and cover soil soon after installation. GCLs installed without an overlying geomembrane and/or on a drier subgrades were much more permeable (Scalia and Benson 2010).

Although significant advances have been made in the knowledgebase regarding changes in soil structure and hydraulic properties, they only represent a snap shot in the evolution of fina covers. These advances will reduce the uncertainty in long-term performance assessments, out additional research investments are needed to more accurately and completely define very long-term structure properties of earthen cover materials corresponding to 100s or 1000s of years. These research investments should include analog studies of natural environments mimicking cover profiles, accelerated laboratory experiments that can be used to develop predictive degradation models, and long-term monitoring of field sites.

Acknowledgement

The Landfill Partnership of the Consortium for Risk Evaluation and Stakeholder Participation (www.cresp.org) and the Wisconsin Distinguished Professorship program provided financial support for Benson's participation in this workshop

References

Albrecht, B. and Benson, C. (2001). Effect of desiccation on compacted natural clays. *Journal of Geotechnical and Geoenvironmental Engineering*, 127(1), 67-75.

Albrecht, B. and Benson, C. (2002). Closure to discussions of "Effect of desiccation on compacted natural clays," *Journal of Geotechnical and Geoenvironmental Engineering*, 128(4), 356-360.

ALBRIGHT, W., BENSON, C., GEE, G., ROESLER, A., ABICHOU, T., APIWANTRAGOON, P., LYLES, B., AND ROCK, S. (2004), FIELD WATER BALANCE OF LANDFILL FINAL COVERS. *J. ENVIRONMENTAL QUALITY*, 33(6), 2317-2332.

Albright, W., Benson, C., Gee, G., Abichou, T., Tyler, S., and Rock, S. (2006). Field performance of three compacted clay landfill covers. *Vadose Zone Journal*, 5(4), 1157-1171.

Albright, W., Benson, C., and Waugh, W. (2010), *Water Balance Covers for Waste Containment: Principles and Practice*, ASCE Press, Reston, VA, 158 p.
Benson, C., Abichou, T., Olson, M., and Bosscher, P. (1995). Winter effects on the hydraulic conductivity of a compacted clay. *Journal of Geotechnical and Geoenvironmental Engineering*, 121(1), 69-79

Benson, C., Albright, W., Fratta, D., Tinjum, J., Kucukkirca, E., Lee, S., Scalia, J., Schlicht, P., and Wang, X. (2010), Engineered Covers for Waste Containment: Changes in Engineering Properties and Implications for Long-Term Performance Assessment, Office of Research, Nuclear Regulatory Commission, Washington, DC.

Benson, C. and Othman, M. (1993). Hydraulic conductivity of compacted clay frozen and thawed in situ. *Journal of Geotechnical and Geoenvironmental Engineering*, 119(2), 276-294.

Benson, C., Sawangsuriya, A., Trzebiatowski, B., and Albright, W. (2007), Post-construction changes in the hydraulic properties of water balance cover soils. *Journal of Geotechnical and Geoenvironmental Engineering*, 133(4), 349-359.

Benson, C., Thorstad, P., Jo, H., and Rock, S. (2007). Hydraulic performance of geosynthetic clay liners in a landfill final cover, *Journal of Geotechnical and Geoenvironmental Engineering,* 133(7), 814-827.

Chamberlain, E. and Gow, A. (1979). Effect of freezing and thawing on permeability and structure of soils. *Engineering Geology*, 13, 73-92.

Dwyer, S. (2001). Finding a better cover. *Civil Engineering* pp. 59-63.

Egloffstein, T. (2001). Natural bentonites-influence of the ion exchange and partial desiccation on permeability and self-healing capacity of bentonites used in GCLs. *Geotextiles and*

Eigenbrod, K. (2003). Self-healing in fractured fine-grained soils. *Canadian Geotechnical Journal*, 40, 435-449.

Jo, H., Katsumi, T., Benson, C., and Edil, T. (2001). Hydraulic conductivity and swelling of nonprehydrated GCLs permeated with single-species salt solutions. *Journal of Geotechnical and Geoenvironmental Engineering*, 127(7), 557-567.

Meer, S. and Benson, C. (2007). Hydraulic conductivity of geosynthetic clay liners exhumed from landfill final covers, *Journal of Geotechnical and Geoenvironmental Engineering*, 133(5), 550-563.

Melchior, S. (1997). In-situ studies of the performance of landfill caps (compacted soil liners, geomembranes, geosynthetic clay liners, capillary barriers). *Land Contamination and Reclamation*, 5, 209-216.

Scalia, J. and Benson, C. (2010), Hydraulic conductivity of geosynthetic clay liners exhumed from landfill final covers with composite barriers, *Journal of Geotechnical and Geoenvironmental Engineering*, in press.

Shackelford, C., Benson, C., Katsumi T., Edil, T., and Lin, L. (2000). Evaluating the hydraulic conductivity of GCLs permeated with non-standard liquids *Geotextiles and Geomembranes*, 18, 133-162.

Suter, G., Luxmoore, R., and Smith, E. (1993). Compacted soil barriers at abandoned landfill sites are likely to fail in the long term. *Journal of Environmental Quality*, 22(2), 217-226.

4.5.3 Assessment of the Erosional Stability of Encapsulation Caps and Covers at the Millennial Timescale: Current Capabilities, Research Issues and Operational Needs

Garry R. Willgoose

Faculty of Engineering and the Built Environment,
The University of Newcastle, Callaghan, 2308, Australia
garry.willgoose@newcastle.edu.au

The Newcastle group have over the last 20 years pioneered the use of landform evolution models (LEM) for the assessment of the long-term stability of encapsulation structures for the mining industry (uranium among them; the wastes have included mine tailings, pyrite and other potentially environmentally sensitive waste) and low level nuclear waste repositories around the world. This work has focused on design lifetimes of 100 to 1000 years, sufficient time for the landforms and ecosystems on them to have changed due to runoff and erosion, so that traditional, fixed landform, erosion models (typically developed for agricultural applications) and remediation solutions are not appropriate. Accordingly we have had to pioneer new ways of (1) calibrating models to data, (2) validating the predictions of the models, (3) methods for analyzing the assessment results, and (4) tools for facilitating the assessment task. In recent years other groups have started using our software or similar tools. This paper will summarize current science and technical capabilities, highlight science challenges that have arisen out of this work that should be resolved, and impediments to the use and acceptance of these approaches in practice.

At their most fundamental level, and at the risk of oversimplifying, LEMs are erosion models where the landform evolves with time in response to that erosion and where the erosion changes in response to the changing landform. For instance, if landform is allowed to evolve for one year then the difference between the landform at the start and end of the year gives the erosion and/or deposition during that year. If the physics modelled are the same, and the amount of landform change in that year is small, then LEMs will (and do) give the same result as traditional erosion models (we use "traditional" in this presentation to refer to models that do not simulate the evolution of the landform; e.g. USLE, CREAMS, WEPP). It is from this starting point that we will discuss the use of LEMs.

All LEMs simulate some form of fluvial erosion model for the hillslopes and channels. They may also model other processes that are crucial once the landform begins evolve (e.g. soil creep, debris flows) and which may or may not be important for any specific application. That said there are some aspects of the parameterisation of erosion physics that are more important for LEMs than they are for traditional models. These will discussed in the context of geomorphic design principles for constructed landforms.

All LEM fluvial erosion models have a sediment transport process that increases in flux/unit width with increasing water discharge/unit width. This means that if flow converges as it moves downstream as, for instance, the result of the development of a rill then the erosion within the rill will increase relative to the adjacent uniform sheetflow. This means that, in the absence of other processes (e.g. rainsplash), rills and gullies develop as result of the positive reinforcement of the physics of fluvial erosion. In principle LEMs can directly model this process where the processes interact to change an initially smooth landform surface into a rilled surface. This rilling and gullying is important for the integrity of covers and caps. Since failure of the structure will occur if any of the waste is released, the capping layer must resist erosion. I will show examples when overland flow convergence occurs as a result of landform evolution and the landscape becomes covered in a network of high erosion regions (the gullies) separated by the low erosion regions (the ridges). Failure occurs when the gullies penetrate the capping. It's of little consequence to know that the average erosion across the whole structure is less than the thickness of the capping if the point of deepest erosion determines failure. The deepest erosion is much higher than the average because the average includes all the intervening ridges where erosion is very low.

The presentation will present some examples of landform assessments going back to the early 1990's using LEMs showing how the LEM assessment can elucidate aspects of the landform stability that more traditional models cannot. These examples are covered in detail in Willgoose and Hancock (2010). These include above-grade structures, geomorphically optimal designs and the potential uses of multi-layered armoured caps and covers.

The discussion above highlights the new insights that can be made using LEMs. However, they are not without their limitations at the current time. At the science level, an evolving landform imposes limitations, which may or may not be important for any particular application. Many things that can be assumed to be constant in a traditional model evolve in concert with the landform. Soils change as the soil armours in the bottoms of gullies. This gully and armour development changes the spatial distribution of the soil erodibility in a way that is intimately linked with the evolution of the landform. Similarly, the soil moisture distribution may also change with the evolving landform. This will change the distribution of vegetation density (particularly in arid regions), and therefore the cover factor, across the landform.

At a practical level, a LEM must solve the erosion equations many millions of times as the landform evolves over time whereas a traditional models only has to solve the erosion equations once. For a LEM this increases the computing load significantly. Some of the LEMs are quite efficient but they will always use more computer time than a traditional model. Small areas (a few square kms at a resolution of 10's of metres, or a million computational nodes in space) can be easily run on a desktop computer with modest run times. Long simulations for large areas can still require significant compute times (CPU days).

The CPU time issue leads to one of the most common criticisms of LEMs from erosion modellers from a traditional background. This criticism is that the physics of LEMs is simpler than in their current model. This is generally in reference to the fluvial erosion model. Much of this criticism is, in our view, misguided. As noted above the fundamental physics in the best LEMS and the most recent process-based traditional models are very similar. Many details such as spatial distribution of soil cover, practice and soil properties are commonly ignored by LEMs but this is a practical issue reflecting their research heritage rather a fundamental flaw. For instance the EAMS erosion assessment package built around the first author's SIBERIA LEM allows the input of these management factors, though at the current time its more of an exception to the rule in that regard.

At the managerial and decision maker level the main limitation to LEM application is that LEMs are still an unfamiliar tool. The LEM user may sometimes address this through education of the manager. In other cases this unfamiliarity can be important for highly political applications. LEMs have not yet been proven in court and this can be important for projects that will need to be legally defended. In these latter applications the LEMs unquestionably provide insight that cannot be obtained any other means. The problem is that if millions of dollars are to be spent and justified using LEM simulations, some assurances are needed that the model's predictions (sometimes hundreds and thousands of year into the future) are, if not correct, at least not potentially misleading. An important related issue is the need to identify indicators based on short-term performance (e.g. 10-20 years of monitoring) that can be used by regulators, as part of rehabilitation sign-off, to provide confidence in long-term (e.g. 100-1000 years) LEM predictions. Progress has been slow in this area, but will be driven by practical applications. These practical applications will elucidate the types of validation required. These validation test cases are qualitatively different from the type of validation needed for traditional erosion models.

The main issue with LEMs is that many hillslope properties that can be directly measured for an existing hillslope actually evolve in concert with an evolving hillslope. This complicates the modelling of an evolving hillslope because this indicates that we need sub-models for the evolution of these hillslope properties and these sub-models must respond dynamically to the evolution of the hillslope. The best example of this need for sub-models is the evolution of the grading of soil on the surface of the hillslope. As a slope or gully evolves erosion strips out the finest fraction of the soil on the surface leaving behind a coarse armour that is relatively more resistant to erosion. As the surface erodes further even more fines are eroded and the surface continues to coarsen. Thus the erodibility of the surface is intimately tied to the cumulative erosion from the surface. To illustrate the scale of this problem consider that for a typical batter on a waste repository edge 20m long and 10m high a change in elevation of 10cm will make only a slight difference to the hillslope gradient and slope (and thus erosion if the erodibility is unchanged), but might make a dramatic difference to armouring of the surface and thus the erodibility of a surface. This suggests that for waste repositories the changes in the surface grading may be as significant, and sometimes more, than the changes in slope. Recent advances (Cohen et al 2010) have overcome this problem, and combined landform and soil evolution models will soon be implemented.

It's still early days in the development of LEMs, particularly with respect to their application to applied problems. At an exploratory level it is relatively easy for a researcher to write a LEM. Accordingly there has been an explosion of research-focussed models in the last 5 years. Many of them have strong underlying similarities in the physics but they typically reflect the interests of the researcher so that there is much that is unique in each particular model. Generally, these research-focussed LEMs do not have the complete set of physics, support and analysis tools, and documentation that are required by the erosion practitioner. Nor, critically, do they have the set of validation tests needed by the practitioner to be able to defend the results of their models. In some cases, like the Monte-Carlo aspects of valley development on landforms (where comparing a landform in the field with a single computer simulation is not possible even in principle), we are only at the early stages of developing testing methodologies (Willgoose et al, 2003).

In particular there is a need for a tool with an easy to use interface, which is well documented, computationally robust, widely validated on the types of problems of interest, contains all the physics that might be needed for typical applications and with a database of erosion parameters (e.g. using pedo-transfer functions). No LEM currently available meets all of these needs. Coulthard (2001) in a recent review, albeit aimed at a research audience, indicated that the author's EAMS-SIBERIA rehabilitation design package most closely met these criteria. In the nine years since little has changed. This reflects the wide range of (mainly) mine rehabilitation case studies on which it has been applied. It is anticipated that a new model, TelluSim, will supersede EAMS-SIBERIA in the near future (Willgoose, 2009). This new model will facilitate new applications in new areas and will make it easier to customise for specific applications and user needs. There are also user needs that are not or only partially met by existing capabilities. Design of covers requires the ability to input spatially distributed layers of materials with selected properties and the ability to track the sources, sinks and travel distances of material as these layers are penetrated, typically only in localised areas. SIBERIA has the capability to model this but the science and model testing is lagging for this application. Some field examples of where this might be important will be shown.

In conclusion landform evolution models (LEMs) are emerging as a practical tool for simulating erosion for a range of problems that are not possible for traditional erosion models (where the landform is fixed and doesn't evolve). Broadly speaking these problems are ones where features that take some time to develop (e.g. gullies), which develop only in response to erosion, and where the features are critical to the success or otherwise of the project. The authors' experience is mostly in the area of mine rehabilitation, waste containment and nuclear waste repositories. In these cases failure occurs when the maximum depth of erosion incision reaches the waste. In these cases it is the maximum depth of incision of gullies and valleys that develop on the landform rather than the average erosion that is critical. LEMs are ideally suited to address this type of problem because of their ability to model the incision process over time. This localised incision is a key part of the development of hills and valleys in landform evolution.

References

Cohen, S., G. R. Willgoose, and G. R Hancock, 2010: A computationally efficient spatially distributed soil pedogenesis model: modelling framework and analysis of hillslope catena. *Journal of Geophysical Research (Surface Processes)*, in press.

Coulthard, T. J. 2001. Landscape evolution models: a software review. *Hydrological Processes* 15:165-173.

Willgoose, G. R. 2009. TELLUSIM: A Python Plug-in Based Computational Framework for Spatially Distributed Environmental and Earth Sciences Modelling. in 18th *World IMACS / MODSIM Congress*, Cairns, Australia.

Willgoose G. R., and G. R. Hancock. 2010. Applications of long-term erosion and landscape evolution models. in *Handbook of Erosion Modelling*. R. P. C. Morgan and M. A. Nearing, (Ed). Wiley-Blackwell Oxford. in press.

Willgoose, G. R., G. R. Hancock, and G. A. Kuczera. 2003. A framework for the quantitative testing of landform evolution models. Pages 195-216 in P. R. Wilcock and R. M. Iverson, editors. *Predictions in geomorphology*. American Geophysical Union, Washington DC.

4.5.4   Ecological Processes and Changes in the Performance of Covers

Steven O. Link

Department of Science and Engineering
Confederated Tribes of the Umatilla Indian Reservation
46411 Timine Way
Pendleton, OR 97801
StevenLink@ctuir.org
stevenlink123@yahoo.com
541-429-7954 (W)
541-969-1159 (C)
509-948-0054 (C)

Ecological processes have caused changes in the performance of covers over the short-term (Waugh, Peterson et al. 1994) and are likely to cause changes in the long-term (Waugh, Peterson et al. 1994). Changes in ecological process are caused by fire (Whisenant 1990; Link, Keeler et al. 2006), erosion, deposition, earthquakes and subsequent deformations, frost heaving, climate change, pedogenesis, and humans (Waugh, Peterson et al. 1994). Ecological processes that can change are associated with plant community dynamics, soil cryptogams, soil microbes, and animals. This review addresses concerns for cover efficacy and suggests research and management topics that may resolve such concerns.

The most significant short-term and long-term ecological degradation processes causing increases in radon release, water percolation, erosion, and bio-uptake are root growth and intrusion, plant composition and cover (Link, Waugh et al. 1994), and animal intrusion and bioturbation of soils (Waugh, Peterson et al. 1994). Increases in radon release have been associated with roots breaking through clay caps. If clay caps are to continue use as radon barriers then regularly scheduled monitoring and maintenance will be required. Such maintenance will include keeping burrowing animals and deep-rooted plants off the cover. Keeping humans off the cap would require significant control measures. Water percolation is primarily affected by plants, but animals can also influence the process (Waugh, Peterson et al. 1994). Lack of sufficient plant cover after fire or other perturbation such as pathological death of a significant number of plants can allow water to accumulate in the top soil layer with possible drainage depending on precipitation levels and soil depth and type. Percolation rates can be affected by root channels and animal burrows that can form a preferential pathway (Devitt and Smith 2002; Albright, Benson et al. 2004; Breshears, Nyhan et al. 2005). Deposition of sands can also lead to increased percolation especially if the plant community is not able to remove the addition water because they are negatively affected by sand (Fayer and Gee 2006). Erosion is strongly controlled by plants (Link, Waugh et al. 1994). If there is insufficient plant cover after a fire for instance, then water and wind erosion are likely (Leary, Berlin et al. 2010). If there is significant erosion after fire then there is a risk of significant soil loss with repeated fires. Gravel admixes in surface layers eventually would protect from wind erosion, but only after significant soil was lost. The effect of a functional layer of pea gravel on soil surfaces on soil water accumulation should be examined. More research is needed to assess plant and root processes plus processes that control preferential flow on caps and relevant analog sites. Research on the effect of fire on plants is needed to assess the how significant a loss of vegetation will be on percolation and erosion. Possible solutions to these issues include institution of regular monitoring and implementation of vegetation management after fire and other disturbances.

Climatological and associated ecological changes can affect the degradation processes. Cover systems are designed with current resources or plants from the local area or areas with similar climates. Climate change will eventually change the conditions for which a cover was designed. If the climate changes then plants on covers will eventually change. If the climate warms and dries then vegetative cover may be reduced increasing risk of wind and water erosion. If climate change results in warmer and wetter conditions then there may be increased risk of water erosion, percolation and drainage. Research to address climate change can be done by comparing a design used in a particular climate with the same design in a warmer and wetter climate. This could be done with lysimeters.

There are ways to minimize degradation processes to reduce radon release, percolation, erosion, and bio-uptake for various ecologies and climates. Making the cover cap thicker will reduce many problems. Thicker soil caps will store more water, accommodate deeper-rooted plants, reduce the impact of animal burrows, and reduce the likelihood of bio-uptake. Maintaining active monitoring and vegetation management will also minimize degradation processes.

Designing changes to reduce one degradation process can cause undesired increases of another. For example, activities that reduce erosion such as using pea gravel to create a desert pavement could increase storage and possible drainage if it becomes continuous on the surface. A way to reduce this risk is to rely on plants to control erosion. An undesired consequence of this strategy would be no control of erosion if plant cover were reduced by repeated fire.

Our understanding of degradation processes can be used to improve the design and performance of covers and liners. After degradation processes have been recognized and understood, research can be commissioned to test possible design improvements. Implementation of new designs then should improve performance of covers and liners.

The key to long-term performance is to recognize that geology, soils, climate, and ecological processes cn covers will change. Such changes can lead to cover failure especially if covers are not maintained. Maintaining cover performance requires periodic inspections, the ability to test new ideas, and then making appropriate modifications. Without maintenance, it is likely that most covers will eventually fail. We can design experiments to improve the performance of short-term caps with respect to vegetation, but cannot use this knowledge to predict long-term performance.

The views of the author may not reflect the policies of the Confederated Tribes of the Umatilla Indian Reservation regarding the use of landfill covers.

References

Albright, W. H., C. H. Benson, et al. (2004). "Field Water Balance of Landfill Final Covers." J. Environ. Qual. **33**: 2317–2332.

Breshears, D. D., J. W. Nyhan, et al. (2005). "Ecohydrology Monitoring and Excavation of Semiarid Landfill
Covers a Decade after Installation." <u>Vadose Zone Journal</u> **4**: 798–810.

Devitt, D. A. and S. D. Smith (2002). "Root channel macropores enhance downward movement of water in a
Mojave Desert ecosystem." <u>Journal of Arid Environments</u> **50**: 99-108.

Fayer, M. J. and G. W. Gee (2006). "Multiple-Year Water Balance of Soil Covers in a Semiarid Setting." <u>J. Environ. Qual.</u> **35**: 366–377.

Leary, K., G. Berlin, et al. (2010). <u>Fire Impacts on a Engineered Barrier's Performance: The Hanford Barrier One Year After a Controlled Burn.</u> WM 2010 Conference Proceedings.

Link, S. O., C. W. Keeler, et al. (2006). "*Bromus tectorum* cover mapping and fire risk." <u>International Journal of Wildland Fire</u> **15**: 113-119.

Link, S. O., W. J. Waugh, et al. (1994). The role of plants on isolation barriers systems. <u>InSitu Remediation: Scientific Basis for Current and Future Technologies, Proceedings of the 33rd Hanford Symposium on Health and Environment,</u>. G. W. Gee and N. R. Wing. Columbus, Battelle Press: 561-592.

Waugh, W. J., K. L. Peterson, et al. (1994). Natural analogs of the long-term performance of engineered covers. In: InSitu Remediation: Scientific Basis for Current and Future Technologies. <u>Proceedings of the 33rd Hanford Symposium on Health and Environment</u>. G. W. G. a. N. R. Wing. Pacific Northwest Laboratory, Richland, Washington Battelle Press, Columbus, Ohio: 379-409.

Whisenant, S. G. (1990). <u>Changing fire frequencies on Idaho's Snake River plains: ecological and management implication</u>, USDA Forest Service Intermountain Research Station General Technical Report: INT-276:4-10.

4.5.5 Degradation Processes and Changes in the Performance of Geomembranes

R. Kerry Rowe

Vice-Principal (Research)
and
Professor of Civil Engineering,
Queen's University, Kingston, Ontario

This abstract (and the presentation) is largely a condensed version of a paper by Rowe (2009) which discusses, amongst other things, the service life of high density polyethylene (HDPE) geomembranes (GMs) used in geoenvironmental applications. The service life may be considered to be the length of time the geomembrane acts as an effective hydraulic and diffusive barrier to contaminant migration (Rowe et al. 2004). Any geomembrane is expected to experience some ageing or degradation during its service life. Typical modes of degradation include oxidation, extraction, biological degradation, ultraviolet (UV) degradation, and thermal degradation. For buried geomembranes not exposed to UV degradation, oxidative degradation is the primary concern for HDPE geomembranes used as covers and bottom liners in most landfills (Hsuan and Koerner 1995; Sangam and Rowe 2002b). Oxidation in the polymer increases exponentially with temperature (Hsuan and Koerner 1995; Sangam and Rowe 2002a). Oxidative degradation causes a breakdown in polymer chains which changes the physical and mechanical properties of the geomembrane and eventually leads to geomembrane embrittlement and failure. The most likely cause of failure of HDPE GMs is considered to be environmental stress cracking of the geomembrane at locations of elevated stress, once the stress crack resistance has dropped sufficiently (Rowe 2009).

There is a paucity of data relating the effect of radionuclide's in low level radioactive waste on the service life of common geomembranes. The effect could be three fold: (a) if the waste is sufficiently active to generate significant heat then that heat will, in and of itself, reduce the service life; (b) if the waste generates a high pH leachate, the pH could affect the GM service life (e.g. by accelerating depletion of phenolic antioxidants); and (c) the radiation from the waste could cause degradation of the GM over and above that which would be expected due to causes (a) and (b). Research is currently in progress with respect to (a) and (b) as discussed below. Some of the issues regarding (c) will then be discussed.

Modern HDPE geomembranes are manufactured from a medium density polyethylene (MDPE) resin to which 2-3% carbon black is added for UV stabilization and it is the carbon black that pushes the geomembrane into the high density (HDPE) classification. Earlier geomembranes were manufactured from true HDPE resin but problems were encountered with these older geomembranes due to inadequate environmental stress crack resistance. There are a number of different resins used by geomembrane manufacturers and these resins may have quite different properties. For example, Scheirs (2009) reports that the two most widely used HDPE geomembrane resin grades (Marlex K306 and K307 from Chevron Phillips) have very different molecular weights due to that fact that they are produced for two different methods of geomembrane production with K306 being used for geomembranes produced using round dies (blown film method) and K307 for geomembranes produced using flat dies (cast extrusion method). Many other resins are available and are used by manufacturers. Even for a given manufacturer the resin used in the geomembrane can vary from time to time. Thus important properties of the geomembrane such as the environmental stress crack resistance (SCR) can vary substantially even for the same manufacturer depending on the resin used at any given time.

Typically at least one primary and one secondary antioxidant is added to the HDPE resin to delay or retard the oxidation reactions both during and following manufacture of the geomembrane. Antioxidants may represent 0.2-0.5% of the geomembrane by weight (Grassie and Scott 1995). Antioxidant packages may change from manufacturer to manufacturer and, for a given manufacturer, from time to time. Hence the service life of one geomembrane will likely not be the same as that for another geomembrane under the same exposure conditions unless the geomembranes have the same thickness, resin and antioxidant package. This should be kept in mind when considering the service lives of geomembranes.

The oxidative degradation of HDPE geomembranes is considered as a three-stage process (Hsuan and Koerner 1998): the depletion of antioxidants (Stage 1), induction time to onset of polymer degradation (Stage 2), and degradation to failure (Stage 3). The service life of a geomembrane is taken as the sum of the duration of the three stages. Stage III is characterized by significant changes to the physical and mechanical properties which will eventually lead to geomembrane failure. Failure in this context refers to a decrease in an engineering property (e.g. stress crack resistance) to a specified value.

Laboratory based accelerated ageing tests are utilized to estimate the length of these stages with the emphasis to date being on Stage I (e.g. Hsuan and Koerner, 1998; Sangam and Rowe, 2002a; Müller and Jacob, 2003; Gulec et al. 2004; and Rowe et al. 2008). These studies have examined extraction of antioxidants from geomembrane immersed in air, water, municipal solid waste leachates, mine drainage and various hydrocarbons. The service life of the geomembrane will depend on a number of factors such as: (a) the geomembrane properties, (b) the immersion fluid, (c) geomembrane thickness, (d) the liner configuration, (e) the protection layer above the geomembrane, (f) applied stress, (g) location in the liner system (base or cover), and (h) most importantly, the time-temperature history of the liner.

There is relatively little data relevant to modern geomembranes for more that Stage I, although tests to address this issue are underway. Long-term performance assessment also require long-term testing (a decade or more) ever when elevated temperatures are used to accelerate aging (Rowe et al. 2009). Also most tests conducted are for geomembranes immersed in the fluid of interest (e.g. landfill leachate) and these tests are typically more severe than would be expected in a real field situation (Rowe and Rimal 2008, Rowe et al 2010).

There is remarkably little data regarding the effect of radiation on the service life of geomembranes. There are anecdotes of studies having been conducted. However, if these studies have been conducted, the information has not found its way to the archival literature. This represents a major void in the literature.

Resent research into the long-term performance of geomembrane liners has indicated that:
- The service life of one geomembrane will likely not be the same as that for another geomembrane under the same exposure conditions unless the geomembranes have the same thickness, resin and antioxidant package. Since the antioxidant packages and resins used by geomembrane manufactures change from time to time, the geomembrane service lives can be expected to change from manufacturer to manufacture and even from time to time with the same manufacturer.

- The constituent MSW of leachate responsible for the significant difference in antioxidant depletion rate between geomembranes immersed in water and synthetic leachate was surfactant (e.g. soap).
- Antioxidants were depleted at a faster rate in relatively acidic or basic immersion mediums (pH 4 and 10) than at typical MSW leachate (pHs 6-8). This suggested that the service lives for geomembranes in contact with MSW leachate may be different to that of geomembranes used for low level nuclear waste if the pH of the leachate generated is high.
- Antioxidant depletion was about 2-5 times faster for a geomembrane immersed in leachate than the same geomembrane in a simulated composite liner. Thus predictions of service life based on immersion tests are likely to be quite conservative
- At temperatures of $20^{\circ}C$ or less, current research suggests that the service life of HDPE GMs used to contain MSW is likely to be of the order of a thousand years. When consideration is given to the likely field situation (as opposed to immersed in leachate) the service life is likely many thousands of years at $20^{\circ}C$ or less but this drops to decades at $60^{\circ}C$.
- Other things being similar, a thicker geomembrane is likely to have a longer service life than the thinner geomembrane.
- There is a paucity of data in the archival literature regarding the effect of low level radiation on the service life of geomembranes. This represents a major void in the literature.

References

Grassie, N. and Scott, G. (1985). *Polymer Degradation and Stabilization,* Cambridge University Press, New York.

Gulec, S. B., Edil, T. B., and Benson, C. H. (2004). Effect of acidic mine drainage on the polymer properties of an HDPE geomembrane. *Geosynt. Int.*, **2**(11): 60-72.

Hsuan, Y. G. and Koerner, R. M. (1995). Long-term Durability of HDPE Geomembrane. Part 1: Depletion of Antioxidant. *GRI Report 16*, Geosynthetic Research Institute, Drexel Univ., Philadelphia.

Hsuan, Y. G. and Koerner, R. M. (1998). Antioxidant depletion lifetime in high density polyethylene geomembranes. *J. Geotech. Geoenv. Eng.*, **124**(6): 532–541.

Müller, W. and Jacob, I. (2003). Oxidative resistance of high density polyethylene geomembranes. *Poly. Degrad. Stab.*, **79**(1): 161–172.

Rowe, R.K. (2009). "Long-term performance of leachate collections systems and geomembrane liners for MSW landfills", Keynote lecture, GeoAfrica 2009, Capetown, September 2009, CD-ROM: 1-23.

Rowe, R.K. and Rimal, S. (2008). "Depletion of antioxidants from an HDPE geomembrane in a composite liner", ASCE *Journal of Geotechnical and Geoenvironmental Engineering*, **134**(1):68-78

Rowe, R.K., Quigley, R.M., Brachman, R.W.I., Booker, J.R. (2004) *Barrier Systems for Waste Disposal Facilities*, Taylor & Francis Books Ltd (E & FN Spon) London, 587p

Rowe, R.K., Islam, M.Z. and Hsuan, Y.G. (2008). "Leachate chemical composition effects on OIT depletion in HDPE geomembranes", *Geosynthetics International*, **15**(2):136-151.

Rowe, R.K., Rimal, S. and Sangam, H.P. (2009). "Ageing of HDPE geomembrane exposed to air, water and leachate at different temperatures", *Geotextiles and Geomembranes*, **27**(2):131-151.

Rowe, R.K., Islam, M.Z., Brachman, R.W.I., Arnepalli, D.N. and Ewais, A. (2010). "Antioxidant depletion from an HDPE geomembrane under simulated landfill conditions", ASCE *Journal of Geotechnical and Geoenvironmental Engineering*, **136**:(7): 930-939.

Sangam, H. P. and Rowe, R. K. (2002a). Effects of exposure conditions on the depletion of antioxidants from high-density polyethylene (HDPE) geomembranes. *Canadian Geotech. J.*, 30, 1221-1230.

Sangam, H. P. and Rowe, R. K. (2002b). Durability of HDPE geomembranes – A Review. *Geotextiles and Geomembranes*, **20**(2): 77–95.

Scheirs, J. (2009). *A Guide to Polymeric Geomembranes: A Practical Approach*, John Wiley & Sons, Ltd, London .

SESSION 3:

EXPERIENCE WITH MONITORING DEVICES AND SYSTEMS USED TO MEASURE PERFORMANCE

Session Chairs:
William Albright, Desert Research Institute/University of Nevada
Craig Benson, University of Wisconsin /CRESP

Technical Reporter:
Robert Johnson, NRC/FSME

5.1   Motivation

Most waste containment facilities require monitoring to verify performance and/or support predictive modeling. However simple in concept, the aims and design of monitoring systems quickly becomes complex as a result of technological challenges and complex goals. Presenters and panelists in Session 3 "Experience with Monitoring Devices and Systems Used to Measure Performance"- of this NRC Workshop were asked to address the following:

- What areas should be monitored for significant degradation/performance (i.e., what are the important processes and components to be monitored)?
- Which barrier systems can be effectively monitored (*in situ* and remotely), and for how long?
- What tools, techniques, and methodologies are available for monitoring, and where/when should they be applied?
- What type and level of monitoring should be done (data sufficiency), and for how long?
- Does monitoring in the short-term provide insights and possible understanding of long-term issues?
- How important are information gaps in monitoring?

Key topics from the presentations are summarized here.

5.2   Background

Monitoring of containment system performance has often been limited to measurements of soil and material hydraulic properties, and the variables related to water status (i.e. soil water content and potential). These limitations are likely to be an outgrowth of the background of the scientists and engineers who were tasked with the design of such facilities, and their lack of knowledge of post-construction changes to system components. There has also been insufficient connection between monitoring and predictive modeling. Accurate prediction is hampered by lack of data to represent both spatial and temporal distributions of system properties and conditions, as well as, processes important to containment system performance. On the other hand, monitoring systems have often not been informed by modeling requirements and the ability of modeling to identify nuances in dynamic systems. Thus, monitoring systems that sometimes do not acquire critical data and modeling results, are often biased due to these insufficiencies in field data.

5.3   Contemporary Issues

It is clear that performance verification of containment systems is important. It is now also clear that understanding the effects of degradation processes is critical to long-term performance. Scientists and engineers face the challenge of understanding and predicting the performance of the entire containment systems, not just individual components. This requires monitoring systems designed to evaluate component interactions and changes to the system that result from environmental influences.

Performance monitoring should include direct measurement of the critical parameters (e.g. percolation through a cover) as well as data from supplemental instruments to understand processes and changes to the containment structure. Too often, the most important performance parameters are calculated from peripheral data or estimated from modeling. Not only is direct measurement critical by itself (when and where possible), but it provides a more accurate and realistic context for interpretation of supplemental data and the status and changes to the containment system.

Monitoring plant processes in particular, and ecological processes in general, can add greatly to understanding cover performance. Even carefully designed cover systems begin a process of change immediately following construction. These changes can affect containment system performance both directly and indirectly, and should be monitored.

Modeling can focus monitoring by identifying key processes and parameters or disconnects between field observations and model results. In return, monitoring can also focus modeling by providing the link between expected and required performance. Thus, monitoring should be driven by the purpose of the project and the modeling objectives to demonstrate performance. Monitoring systems should be designed to understand processes and identify precursors to performance problems.

Remote sensing can provide an alternate view of a containment system and its performance. Advantages include: a more complete spatial coverage of the system; detection of factors not seen by the eye; automation, and potentially lower surveillance costs. Remote sensing can focus evaluations by flagging issues that might have been missed by visual surveillance.

5.4   Recommendations

Several recommendations, both specific and general, were identified by panel members and workshop participants. The topics include:

- The deep vadose zone (DVZ) is a difficult environment to monitor. However, DVZ monitoring is needed. Data from DVZ monitoring are scarce and are typically point measurements. Development of better monitoring system designs and methods should be a priority.
- Concentration data alone do not address transport issues and are not sufficient to develop a comprehensive understanding of overall system performance. Flux protocols and criteria should be developed to focus on overall system performance.
- Develop methods to upscale data on vegetative processes (e.g., evapotranspiration and plant root uptake) to the landscape scale.
- Monitor the entire system. Monitoring and evaluating all components (i.e., cover, tailings, vadose zone below tailings) of a uranium mill tailings site would provide insights into how the containment system works and its efficiency. Monitoring at the regulatory boundary using perimeter wells does not identify precursors of problems prior to a release and does not identify critical components.
- Always consider how uncertainty in monitoring data is propagated through the system performance assessment.
- Consider double liner systems with intervening monitored detection zones between the liners.
- Identify and develop vegetative indicators of system performance.
- Public involvement should include explaining the purpose of monitoring and distinguishing compliance monitoring from parameter and interaction monitoring to improve public communication and understanding.

5.5   Extended Abstracts

The workshop organizing committee considered the topic of monitoring to be a central issue in demonstrating engineered barrier performance. Following the review of technical literature on monitoring of engineered barrier performance, the workshop organizing committee discussed relevant ongoing field monitoring studies. Invited speakers from industry, Federal agencies and academia provided the following extended abstracts. These papers focus on practical insights for developing and implementing monitoring strategies including techniques and instrumentation to confirm short- term performance of the engineered barriers systems.

5.5.1 In Search of the Perfect Cap: 15 Years of Performance Data from the Prototype Hanford Barrier

A. L. Ward

Pacific Northwest National Laboratory
P.O. Box 999 MSIN K9-33, Richland, WA 99352

Engineered surface barriers are recognized as a remedial alternative to the removal, treatment and disposal of near-surface contaminants at a variety of waste sites within the DOE complex. However, stakeholder acceptance of this technology has been negatively impacted by the limited amount of performance data and the uncertainty associated with using these data in support of long-term performance predictions. The culmination of research into the design and performance of engineered barriers, initiated at Hanford in the mid 1980's, was the construction of a field-scale prototype Hanford barrier in 1994. Data collected from this cover are being used to narrow knowledge gaps about constructability and performance.

This prototype Hanford barrier is a 2-ha is a 4.5-m thick multi-component design incorporating a 1-m thick silt loam surface layer with 15% pea gravel to control erosion as well as a capillary break, an asphaltic concrete layer at the base, and two protective side-slope configurations. The cover was designed to meet a 0.5 mm/yr drainage criterion and is instrumented to monitor stability as well water balance components. A treatability test conducted from 1994-1998 included irrigation at a rate of 480 mm/yr including a simulated 1000-yr return storm each March in which 68 mm of water was applied over an 8 hr period. Monitoring has been almost continuous for the last 16 yrs and has focused barrier stability, vegetative cover, evidence of plant and animal intrusion, and the main components of the water balance, including precipitation, runoff, storage, drainage, and deep percolation. The total precipitation received from October 1994 through August 2008 was 3311 mm on the northern half (formerly irrigated), and 2638 mm on the southern, non-irrigated half. Water storage in the fine-soil layer shows a cyclic pattern, increasing in the winter and decreasing in the spring and summer to a lower limit of around 100 mm, regardless of precipitation, in response to evapotranspiration. Topographic surveys show the barrier and side slopes to be stable and the pea-gravel admix has proven effective in minimizing erosion through the creation of a desert pavement during deflationary periods. Only three small runoff events have been observed and the 600-mm design storage capacity has never been exceeded. Total percolation ranged from near zero amounts under the soil-covered plots to over 600 mm under the side slopes. The asphaltic concrete prevented any of this water from reaching the buried waste thereby eliminating the driving force for the contaminant remobilization.

Plant surveys show a relatively high coverage of native plants still persists after the initial revegetation although the number of species decreased from 35 in 1994 to 10 in 2009. There is ample evidence of insect and small mammal use suggesting that the barrier is beginning to function like a recovering ecosystem. Nonetheless, a critical unknown for long-term performance is the effect of wild fire during a post-institutional control environment where routine maintenance may be limited or non-existent. In September 2008, a controlled burn was conducted on one half of the barrier. The effects this sever disturbance on barrier performance have been monitored and documented 2 years. Soil physical, chemical, and hydrologic properties; plant floristics and density; and animal-use were characterized before and after the fire with the unburned half of the barrier serving as a control. Temperatures during the controlled burn ranged from 250 $^{\circ}$C 1.5 cm below the surface to over 700 $^{\circ}$C, 1 m above the surface. Significant decreases in hydraulic conductivity and surface-soil wettability were observed immediately after the fire. Post-fire concentrations of major soil nutrients, pH, and electrical conductivity remain elevated. Dense stands of sagebrush were destroyed allowing many more species to emerge, thereby increasing species diversity. Seed sources contributing to this species diversification were from either the existing seedbank and/or wind-blown sources. There were significant differences in the rate of accumulation and loss of soil moisture on the burned and unburned sections. On the burned section, water storage was higher during the fall; increased more slowly with the onset of winter precipitation (owing to higher evaporation); and decreased more slowly in the spring (owing to lower evapotranspiration). Significant differences in storage between the burned and unburned sections are evident. The dataset generated by this study is unique and have already proven useful in the calibration of barrier design tools (e.g. STOMP) and the development of more rigorous methods for evaluating long-term performance and quantifying associated risk and uncertainty. Results will contribute to a better understanding of long-term barrier performance and the recovery of engineered ecosystems after major disturbances in a post-institutional control environment. Such an understanding is needed to enhance stakeholder acceptance regarding the long-term efficacy of engineered barriers.

5.5.2   ACAP: Monitoring Cover Performance and Changes in Performance with Drainage Lysimeters, Instruments and Exhumations

**William H. Albright[1] and
Craig H. Benson[2]**

[1]Desert Research Institute, Reno, NV 89512;
[2]University of Wisconsin, Madison, WI 53706

Earthen covers relying on water-storage principles (water balance designs) and conventional covers employing resistive barriers were monitored in large instrumented drainage lysimeters over a range of climates at 12 field sites in 8 (US) states as part of the USEPA's Alternative Cover Assessment Program (ACAP). Average annual percolation rates for the conventional covers with composite barriers (geomembrane over fine soil) typically were less than 3 mm/yr. Average percolation rates for conventional covers with soil barriers were between 7 and 160 mm/yr and depended on climate. Average percolation rates for water balance covers ranged between 0 and 110 mm/yr and depended on climate.

The ACAP monitoring system combined a direct measurement of cover performance (the lysimeter) and data about the soil water status (soil water content and suction sensors). This allowed interpretation of mechanisms important to cover function in the context of measured performance. Two examples: preferential flow through the compacted clay covers was suggested by observations of percolation immediately following precipitation events without increased water content in the soil barrier and; drainage through the composite covers was typically accompanied by lateral flow on the barrier membrane and near-saturation of the surface soil layer.

All of the ACAP sites were exhumed following several (4-8) years of field service. Large undisturbed samples were taken from all soil layers, analyzed in the lab for saturated and unsaturated hydraulic properties and compared to similar samples taken during construction. These hydraulic properties, critical to cover performance, exhibited significant change during the service life. Results from the ACAP exhumations provided a preliminary indication of the direction and magnitude of change in hydraulic properties that can be used to adjust these parameters during the design stage.

The combination of direct measurement of cover performance, supporting data from instruments and site evaluations, and in-service soil sampling gives a model for a site investigation in support of future cover design.

References

Albright, W.H. and C.H. Benson (2003). Alternative Cover Assessment Program 2002 annual report. DRI Publication #41182, Desert Research Institute, University and Community College System of Nevada, Reno NV.

Albright, W.H. and C.H. Benson (2004). Alternative Cover Assessment Program 2003 annual report. DRI Publication #41193, Desert Research Institute, University and Community College System of Nevada, Reno NV.

Albright, W.H., C.H. Benson, G.W. Gee, T. Abichou, A.C. Roesler, and S.A. Rock. (2003), Evaluating the alternatives, Civil Engineering, 73(1), 70-75.

Benson, C.H. (2001), Waste containment: Strategies and performance, Australian Geomechanics, 36(4), 1-25.

Benson, C.H., T. Abichou, X. Wang, G.W. Gee, W.H. Albright. 1999. Test section installation instructions, assessment program. Geo Engineering Report No. 99-3. Geo Engineering Program, University of Wisconsin-Madison. Madison, WI, USA.

Benson, C.H., T. Abichou, W.H. Albright, G.W. Gee, and A.C. Roesler. 2001. Field evaluation of alternative earthen final covers, International J. of Phytoremediation, 3(1), 1-21.

5.5.3   Monitoring Contaminant Strategies: Tools, Techniques, Methodologies and Modeling Approaches

Timothy J. Gish[1], Andrey K. Guber[2] and Yakov A. Pachepsky[2]

[1]USDA-ARS, Hydrology and Remote Sensing Laboratory, Beltsville, MD 20705 and the
[2]USDA-ARS Environmental Microbial and Food Safety Laboratory Beltsville, MD 20705

A century-long history of experiments on solute transport in soils has resulted in a wide range of experimental setups and procedures, as well as methods for interpreting observations which has led to considerable ambiguity regarding monitoring approaches. This presentation will focus on results and lessons learned from three shallow chemical transport studies carried out at the USDA-ARS Beltsville, Maryland research facility. In each of the three studies the chemical application area had a radius of several meters to accommodate the scale of possible releases of contaminants from waste storage facilities.

Although no single monitoring setup accurately quantifies chemical transport, future monitoring exercises will likely consist of combining several approaches. There are three general types of setups for monitoring solute transport: 1) outflow breakthrough sampling; 2) destructive sampling after transport has occurred; and 3) monitoring of pore solution concentrations during the transport investigation. Outflow breakthrough sampling is probably the most reliable approach since it effectively integrates solute transport for a given area or volume of soil. However, if little or no chemical breakthrough occurs due to absorption or degradation this method has limited applicability. Furthermore, it is difficult to interpret breakthrough data when several transport processes, retention, and/or metabolic processes occur simultaneously since the impact and interaction of each process must be described from the same single breakthrough curve. Destructive post-experiment sampling is a useful complement to breakthrough data in that it may diminish ambiguity in transport parameterization. However, as the scale of observation coarsen, resource demand increases and field variability increasingly compromises data reliability and interpretation. Monitoring the pore solution composition is in principle the best approach for obtaining transport data. Unfortunately, monitoring devices tend to distort the velocity field within the soil and it is not known from what part of the pore-network system the solution is being extracted. Although the ambiguity of data and/or their analysis creates difficulties in solute transport experiments at the field-scale, the importance of the vadose zone in controlling chemical fate and transport in the subsurface is critical and will likely require the simultaneous use of all three approaches to quantify chemical transport through soil.

An effective monitoring setup will likely involve a hybrid approach where surveys, various models, and solute monitoring experiments interact to better understand chemical transport. For example, subsurface flow and transport modeling can have various purposes including site characterization, explaining the existing subsurface contaminant patterns, and projecting the contaminant release results under various environmental and management scenarios. Soil and geologic surveys can be used to identify critical soil properties which would be used as input into pedotransfer function models that in turn would generate soil water retention and conductivity parameters (Pachepsky and Rawls, 2004; Guber et al., 2009). The pedotransfer function analysis would then be used with geophysical tools like electric resistivity and ground-penetration radar to generate information on pedon-scale heterogeneities (Gish et al., 2002; Newman et al., 2003; Rubin and Hubbard, 2005). The geophysical analysis could reveal the existence of hydrologically active zones as well as delineation structural units that have distinctively different hydrologic behavior. With a basic understanding of site heterogeneity, chemical transport simulations would then be run to determine placement of monitoring instrumentation. Once the monitoring system locations are determined and installed, tracer experiments could be conducted to simulate contaminant leaks and to identify soil regions of high mobility and/or soil features that could influence contaminant transport.

This hybrid monitoring and model approach was tested at the USDA-ARS Beltsville facility and used to demonstrate its applicability for quantifying potential and actual contaminant release from sites germane to US NRC operations. Soil moisture, soil water potential, conservative tracer concentrations in groundwater, groundwater levels, and weather data, along with ground-penetration radar surveys, electric resistivity monitoring, and dilution tests complemented borehole log data and laboratory hydraulic measurements to characterize soil heterogeneity. The first experiment focused on quantifying solute dispersion in unsaturated soil and to evaluate prevailing heterogeneities (Gish and Kung, 2007; Yakirevich et al., 2010). In this study a tracer was applied to a relatively small area around a pumping well, with the irrigated region being substantially larger than the tracer application area. Groundwater was pumped from the well while being continuously monitored for water volumes and tracer concentrations. The second and third studies were meant to evaluate the potential contribution of the variably saturated zone and capillary fringe to lateral chemical transport. The lateral flow and transport processes were found to be complex since the perched groundwater table fluctuated substantially over time. The tracer area was continuously irrigated generating a gradually expanding plume where the groundwater composition during the experiment was monitored using observation wells located within the projected plume. Like the monitoring procedure, no single chemical transport model, regardless of complexity can describe field-scale transport. In this regard model abstract techniques were used with a sequence of model simplifications based on the HYDRUS software family to evaluate chemical transit times (Pachepsky, 2006). The invoked series of model abstractions showed the important role of subsurface heterogeneity in the vadose zone and groundwater, and substantial improved the conceptualization of the subsurface and chemical transport behavior. For this site, data and model analysis indicated that field experimental setups should address at least two types of heterogeneities: 1) differences in water mobility in different parts of the soil; and 2) the effects of restricting layers and the capillary fringe on lateral transport.

References

Gish, T. J., Dulaney, E.P., Kung, K.-J.S., Daughtry, C.S.T., Doolittle, J.A., and Miller, P.T. 2002. Evaluating use of ground-penetrating radar for identifying subsurface flow pathway. Soil Sci. Soc. Am. J. 66:1620-1629.

Gish, T.J., and Kung, K.-J.S. 2007. Procedure for quantifying a solute flux to a shallow perched water table. Geoderma 138:57-64.

Guber, A. K., Pachepsky, Ya. A., van Genuchten, M. Th., Simunek, J., Jacques, D., Nemes, A., Nicholson, T. J., Cady, R. E. 2009. Multimodel simulation of water flow in a field soil using pedotransfer functions. Vadose Zone J 8:1–10.

Neuman, S. P., P. J. Wierenga, and T. J. Nicholson. 2003. A comprehensive strategy of hydrogeologic modeling and uncertainty analysis for nuclear facilities and sites, NUREG/CR 6805. U. S. Nuclear Regulatory Commission. Washington, D.C. 20555-0001.

Pachepsky, Y.A., A.K. Guber, M.T. Van Genuchten, T.J. Nicholson , R.E. Cady, J. Simunek, M.G. Schaap, 2006. Model abstraction techniques for soil-water flow and transport, 165 NUREG/CR-6884, U.S. Nuclear Regulatory Commission Washington, DC 20555- 0001.

Pachepsky, Y. A. and Rawls, W. J. (eds.). 2004. Development of pedotransfer functions in soil hydrology. Elsevier, Amsterdam.

Rubin, Y., and S. S. Hubbard (eds.). 2005. Hydrogeophysics. Springer.

Yakirevich, A., T. J. Gish, J. Šimůnek, M. Th. van Genuchten, Y. A. Pachepsky, T.J. Nicholson, and R. E. Cady. 2010. Potential impact of a seepage face on solute transport to a pumping well. Vadose Zone. IN PRESS.

5.5.4   Aerial Remote Sensing as a Component of Closure Cap Monitoring

John B. Gladden

Savannah River National Laboratory

The problem of near and long term monitoring of closure caps is a vexing one that warrants a review of the underlying objectives of the monitoring program. While the objective of the capping systems to permanently isolate hazardous materials from the environment is clear, the adequacy of the strategies currently in use to ensure that the objective is achieved is less clear. The core of current monitoring strategies involves two components:
- visual and spot topographic surveys of the cap surface, and
- perimeter groundwater monitoring.

More recent capping projects may have more sophisticated schemes such as installed leachate systems that can provide more current information on the stability of the contained waste materials. Installed sensors are also utilized in some facilities. In all cases, these strategies require 'boots on the ground' at some interval, usually no less than annually and frequently quarterly. With widely dispersed and remote disposal sites, such as occur in the west, these strategies represent a substantial long term commitment of manpower and funding over the very long time periods during which the monitoring program must be maintained.

Information provided by the two core monitoring components is relatively unsatisfying for other reasons. Visual inspections are highly dependent on the training and experience of the inspector and will most likely provide only partial coverage of the facility. Topographic surveys of installed subsidence plates provide information on a minute portion of the facility. Installed sensors, with or without telemetry, can provide more detailed information for a variety of parameters, but over a long time period require maintenance and/or replacement, and still only provide information in the immediate vicinity of the sensor probe.

Perimeter wells provide an even less satisfactory solution to ensuring that hazardous materials do not escape from the disposal facility, posing a potential risk to the public and environment. By the time contaminants reach a perimeter well, two undesirable things have occurred:
- key components of the containment structure have failed, and
- there is a groundwater contamination issue that must be remediated.

An alternative monitoring strategy would be based on determination of predictors and early indicators of containment failure. Such a strategy must be based on a comprehensive understanding of the life cycle of a capping system and the potential failure mechanisms that could occur throughout that life cycle. For most capping systems, the key element to maintaining the design function is the proper management of water movement. Penetration of water into the waste mass is the occurrence that the capping system is designed to prevent, and can ultimately result in the release of contaminants. Consequently, approaches that detect abnormal patterns of moisture distribution across the capping system can provide an early indicator of a future loss of containment failure allowing early, and presumably less expensive, intervention.

Vegetative covers on closure caps provide two main functions, stabilizing soils and/or accelerating moisture removal from the surface soil layer through evapotranspiraton. In either case, compromise of the vegetative layer presents a problem for the functioning of the containment system. In addition, the root structures of the plants provides a system that continually samples soil moisture conditions throughout the volume of the root structure which will normally be several fold larger than the above ground plant volume. Changes in water availability for plant roots results in physiological and structural changes in plant leaves which can be detected using a variety of ground level and remote methodologies.

For ET caps the species composition and biomass of the vegetative cover is of critical importance because plants are the pump that removes water which has accumulated during the wet season. For both ET and more conventional capping systems, changes in plant species composition over time can affect the functional performance of the system and result in challenges to other components of the structure. Consequently, approaches that effectively monitor the composition and condition of the vegetative layer may provide early indicators of functional problems in the capping system.

Aerial remote sensing technologies have been used in a wide variety of characterization and monitoring applications such as natural resource management, agriculture and mineral exploration. The suite of technologies is mature and continues to improve in capability as costs are reduced. Whether from aircraft or satellite, sampling intensity is scalable both in space (i.e. larger or smaller sampling units) and time (i.e. measurement frequency) and the variety of sensors allows for a wide range of measurements to be collected.

One such technology, Hyperspectral Imaging (HSI) has been evaluated at the Savannah River Site (South Carolina) and the Monitcello Disposal Cell (Utah) for evaluating the composition and condition of vegetative covers. Analysis of the spectral signatures of vegetation reveals unique signatures that are associated with individual plant species that occur on the caps (Figure 1). Other analysis algorithms relate spectral signatures to the physiological status of the plants that can be associated with moisture stress or other plant stress conditions. Because these are spatially distributed data, abnormal spectral signatures in an individual pixel or cluster can be related to a specific point on the ground, so that inspections by trained personnel can be targeted to specific problem areas.

Figure 1. Hyperspectral Image analysis of Monticello, UT Disposal Cell in 2002 (left) and 2008 (right) showing changes in species composition over an eight year interval using high resolution imagery.

Other technologies, such as Light Distance and Ranging (LiDAR), have also been tested and provide relatively inexpensive, but high resolution, information on changes in topography that could be related to subsidence or erosion. Measurements of vertical changes in the surface layer can be resolved to centimeter levels of resolution. Time series measurements can therefore detect minor changes in surface elevations due to erosion or subsidence. This technology is currently widely used for other applications and requires no further development.

Research needs in the area of spectral analysis include development of stronger relationships between the vegetation responses that can be detected by the remote imaging systems and processes within the capping system that are precursors to, or associated with, deviations from design performance. The most rapid advances are likely to be achieved by reviewing of research and commercial applications conducted in support of agricultural management where spectral analysis is currently used for management of crops.

In summary, the costs of long term monitoring of closure caps represents a significant cost element for long term management of land disposal of wastes and failure to detect loss of waste containment can result in significant, unanticipated costs. Techniques and technologies that are both relatively inexpensive and provide early indications of divergence from the expected behavior of closure caps can improve the protection of public and environmental health, while reducing the cost of both monitoring and potential remedial actions.

5.5.5 Differential Settlement and its Importance on the Performance of Cover Systems at Radiological Waste Disposal Facilities

Robert C. Bachus, Ph.D., P.E., and
John F. Beech, Ph.D., P.E.

Geosyntec Consultants, 1225 Roberts Blvd., Suite 200, Kennesaw, GA

The final cover systems for many old radiological waste disposal sites often represent the last line of protection with regards to minimizing the infiltration of precipitation into the disposed waste. It is, therefore, important that the final cover systems at these facilities maintain their long-term integrity to maximize the protective benefits. For old sites, cover system integrity can be adversely impacted by the historic waste and cover placement practices at the site. For new sites, it is equally important to provide long-term protection to the buried waste. At these facilities, however, owners and operators have the distinct benefit of being able to control the placement of waste and cover materials during active landfilling operations. At both new and old facilities, the integrity of the final cover system is often controlled by differential settlement of the final cover system components. This presentation will focus on two important aspects regarding differential settlement of final cover systems: (i) mechanisms for initiating and inducing differential settlement; and (ii) techniques for predicting and controlling differential settlement. Factors that contribute to differential settlement in radiological waste include: differential stiffness, raveling, and internal erosion. Soils and geosynthetic materials used in constructed final cover systems respond to differential settlement by physical movement, arching, and mobilizing tensile resistance. The roles of these mechanisms and factors, as well as how to calculate (and in some cases control) the relative importance of each will be discussed. This portion of the discussion will focus on how differential settlement may (or may not) impact the hydraulic and physical integrity of the final cover system. With regards to the techniques for predicting and controlling final cover system performance, the presentation will provide a discussion of typical specifications for newly constructed facilities. This portion of the discussion will also identify monitoring procedures and techniques that can be used at both new and old facilities to assess differential settlement.

SESSION 4:

MODELING EXPERIENCES IN PERFORMANCE ASSESSMENT AND EVALUATION OF PERFORMANCE MONITORING

Session Chairs:
David Esh, NRC/FSME and
Thomas Nicholson, NRC/RES

Technical Reporter:
Christopher Grossman, NRC/FSME

6.1   Motivation

Because of the extended timeframes involved with isolating radioactive materials from the environment, the future performance of engineered barriers (e.g. engineered covers and liners) must be estimated in prospective performance assessments of waste disposal and other similar systems. Modeling is a commonly employed tool to estimate future performance. Modeling results may be verified through performance monitoring. Without adequate guidance, modeling may be incomplete and use inconsistent or disparate approaches and untested assumptions. The primary goal of this session was to develop an understanding of the state of modeling of engineered surface barriers and waste liners.

6.2   Background

Numerical modeling has experienced increased acceptance in the facilitation of decision-making, especially with regard to long-term waste management decisions. Increasingly, more complex models reflecting more complex problems are being assessed with numerical models. Modelers are faced with a variety of decisions, but guidance to help facilitate that decision-making and to critically evaluate the models support those decisions is distributed over a large number of information sources.

A variety of topics relevant to the modeling of surface covers and waste disposal liners were discussed. Topics considered in the session included: (a) water balance modeling, incorporation of environmental conditions; (b) changes to materials and system components over time; (c) issues of temporal and spatial scales in measurement and modeling; (d) estimation of water drainage below the root zone (percolation) and radon fluxes; (e) simulation of vegetation dynamics (above- and below-ground); (f) interrelated hydrologic processes (e.g., surface erosion, evapotranspiration (ET), infiltration, percolation, subsurface drainage); (g) the effects of natural and anthropogenic processes and events (e.g., fire, grazing, climate change) on the landscape at spatial scales ranging from less than 1 m^2 to watershed-level (>100,000 hectares) and at time scales ranging from days to centuries or longer; and(h) time periods for these model evaluations.

Session participants were posed a variety of questions to help frame the discussion (see Appendix A Workshop Agenda). Session participants presented information ranging from high-level considerations important to all types of modeling to specific applications such as long-term plant succession modeling. Speakers included John Tauxe (Neptune and Company), Craig Benson (University of Wisconsin), Roger Seitz (Savannah River National Laboratory), Andy Ward (Pacific Northwest National Laboratory), Terry McLendon (KS2 Ecological Services Specialists, LLC), and Bill Kustas (U.S. Department of Agriculture). The panel also included Robert Holt (University of Mississippi), Gary Willgoose, University of Newcastle, Callaghan, Australia, and Ming Zu (DOE).

6.3   Challenges and Experiences in Performance Modeling

Workshop participants provided valuable insights and experiences with respect to the modeling of surface barriers and waste disposal liners. For waste disposal systems, integration of processes in modeling is essential because the processes are integrated in the real world. Influence diagrams and conceptual drawings can be useful in refining the scope of modeling. Graded and iterative approaches, common to performance assessments, have been found to be useful in the modeling of engineered covers and liners. Processes determining cover performance can be non-linear, and typically involve large uncertainties, especially for longer-term assessments. Therefore, integration is extremely important to engineered cover modeling, which needs to take into account these uncertainties in performance predictions. Since the models often include different forcing functions, they need the capability to incorporate the influences of those forcing functions in performance assessments. Model completeness is extremely difficult to assess. At a minimum, models must include water flow and transport, soil loss by erosion, surface-water runoff, and bioturbation by animals and plants. Model validation[1] strategies are useful in assessing the completeness of the models. However, most modeling to date has limited validation due to overly simplistic approaches and the lack of site-specific monitoring data. Calibration exercises are frequently employed, but model confirmation efforts are used rather infrequently. Inter-code comparisons may be useful in understanding code limitations. Probabilistic modeling has been found to be useful to incorporate uncertainties and to identify ranges in performance conditions.

Data should be collected to confirm as-built properties and confirm post-closure performance. Data is generally available to parameterize the initial conditions for the models, although information is somewhat lacking for some geochemical modeling, biotic transport processes and long-term processes. Although short-term data is generally available, information on site-specific processes at high-frequency resolution is rarely collected. High-frequency data may be useful in providing support for systems modeling. High resolution data may be obtained from thermal infrared sensing and other similar techniques.

It is difficult to formulate scenarios for the modeling of engineered covers and liners over long time frames. Participants identified processes, especially hydrologic, that require high frequency data (e.g. daily or hourly) in order to capture observed responses. In addition, simulations of long-term performance must incorporate a variety of potential degradation mechanisms such as extreme weather events, recharge due to flooding, freeze-thaw cycling, fire, and biointrusion. Modeling coupled processes with substantially different response times can be computationally and intellectually challenging, but progress is being made in multi-scale, multi-physics modeling.

Ecological modeling has been progressing, and is a very important component to assessing the performance of engineered covers. For example, roots penetrate natural barriers, and plant succession occurs everywhere. Two common aspects of this succession are an increase in vegetation structure, and in the relative amounts of woody plants. Both of these aspects have implications to the performance of engineered barriers. There is a need to improve the translation of plant impacts to changes in model performance over time corresponding to plant succession periods. Application of the EDYS model, with associated confirmation studies, was presented to workshop participants. Ecological modeling needs to consider a variety of processes, such as but not limited to wind throw of trees potentially perturbing soil layers and pooling runoff, changes to soil microbial communities, and variation of erosion rates with vegetation cover. Of primary importance are the lifeform and species of plants and the characteristics of the materials overlying the barrier.

It is important to collect monitoring and performance validation information and then to couple that information with the modeling activities. Sometimes performance indicators can be identified through the iterative performance assessment approach, and those performance indicators can be measured in the monitoring program. The integration of modeling and monitoring should be systematic and provide indicators to evaluate overall system performance.

---

[1] NRC staff prefers the term 'model confirmation' over 'model validation' since validation is usually not possible due to the long timeframes of predicted performance.

6.4   Recommendations

Recommendations were identified from the information provided by workshop participants. The recommendations were:

- A graded and iterative approach to modeling of engineered covers and liners should be used. In many cases, data is collected during and after engineered barrier implementation. That data is useful to refine and improve the models.

- System-level modeling is useful for predicting engineered cover and liner performance especially for long-time frames. System-level modeling may allow a more complete representation of all of the processes and their couplings even if it means using abstracted, or simplified representations of the system and processes.

- Engineered cover and liner performance modeling must reasonably integrate and couple processes at different time and spatial scales. Many near-surface processes are tightly coupled. Elimination of those couplings may result in incomplete or inadequate conceptual models resulting in errors of predicted performance.

- The level of detail needed in a particular model is site- and application-dependent. Models must be evaluated on the basis of their particular application and not relative to a different application.

- The creation of compilations of relevant data would improve modeling by assisting in more representative parameter estimation and a more thorough consideration of site conditions. In turn, the data would provide development of more realistic conceptual models of engineered covers and liners.

- Model confirmation is essential. It should be a necessary step in performance modeling of engineered covers and liners.

- Modeling activities need to be better integrated with development of monitoring systems to make more effective use of data collected by monitoring systems.

- Long-term modeling should include scenarios based upon plant succession and paleohydrologic evidence.

6.5   Extended Abstracts

The workshop organizing committee considered the topic of modeling to be a central issue in understanding engineered barrier performance. Following the review of technical literature on modeling of engineered barrier performance and overall performance assessment of waste facilities, the workshop organizing committee discussed relevant modeling exercises. Invited speakers from industry, DOE national laboratories and academia provided the following extended abstracts. These papers focus on modeling experiences in performance assessment and evaluating performance monitoring.

6.5.1 Development of an Integrated Probabilistic Model of Radiological Fate and Transport in an Engineered Cover

John Tauxe, PhD, PE

Neptune and Company, Inc.
Los Alamos, New Mexico, USA
jtauxe@neptuneinc.org

Introduction

In radioactive waste performance assessment (PA) and similar environmental contaminant transport modeling work, one must address processes occurring in the materials covering the wastes, since there are upward pathways for contaminant migration. This also applies to any sort of waste, including uranium mill tailings and decommissioned sites that have left radioactive materials in place. Unfortunately, these upward contaminant transport pathways are often ignored, in part due to the historically myopic attention to hydrology, and in part due to the difficulty of characterizing and modeling these pathways.

The discipline of PA arose from the fields of groundwater hydraulics, so the focus on water as a transport medium is understandable. Water transport, however, had generally been assumed to occur in a static system that indefinitely retains the physical properties of its design and construction. The focus of modeling their hydraulic performance was to inform that design, rather than its future evolution in nature. This allowed straightforward modeling using a number of specialized cover modeling programs, such as HELP, UNSAT-H, or VS2DI (USACE 1997, Fayer and Jones 1990, and Hsieh, et al. 2000). In reality, however, we know that even carefully constructed covers do not outlast the wastes below (Suter, et al., 1993). Nevertheless, the processes that lead to their degradation are not traditionally modeled, even though these same process, by their nature, may themselves be or contribute to significant contaminant transport pathways.

The Need for Cover Modeling

Aside from aesthetics, the principal reason for constructing covers in the first place is for containment. In general, we think of waste facilities (in the broad sense) as being areally extensive, with a distinct top and bottom. These may or may not have engineered liners, but in all cases there is some sort of cover. The purpose of the cover is two-fold: to keep waste constituents (contaminants) from moving up to the surface soils and the atmosphere, and to keep water from infiltrating into the wastes and down into the subsurface and groundwater.

There are, of course, many variations on this theme. Extremely arid sites may experience no net downward flow, with water actually migrating slowly upward to the ground surface in response to the substantial evaporation potential, like at the Nevada Test Site (DOE 2006). Other sites may be so wet as to have the waste itself reside in the saturated zone, where groundwater can flow laterally beneath the cover (e.g. at the Savannah River Site tank farms). In between these extremes, the wastes may get periodically "washed" by a variable water table, alternately saturating and draining the waste, like the bathtubbing trenches at Oak Ridge National Laboratory (ORNL 1997). The case we like to think about has a nice impervious cover that sheds infiltrating water like a roof over the waste, keeping it dry and preventing excessive leaching of contaminants into groundwater.

Out interest in the potential for migration of contaminants is motivated by our desire to estimate the risks they pose to human health and the environment, and by the need to optimize the reduction of these risks in the interest of keeping them as low as reasonably achievable (ALARA). While the ALARA concept was developed in the context of radiological exposures, it can be applied to performing cost/benefit analyses for all types of risk-inducing phenomena. So, beyond just determining if they are adequate for meeting some regulatory compliance, we have an interest in understanding their performance as realistically as possible.

The Need for Integrated Models

The real world as not as neat and clean as any single type of model assumes, of course. The processes that influence cover degradation and evolution are diverse. If we constrain ourselves to the point of view of the cover materials, we could include

- waterborne porous medium transport (advection and diffusion)
- airborne porous medium transport (advection and diffusion)
- soil loss due to erosion
- bioturbation by burrowing animals
- plant uptake and redistribution of contaminants
- radioactive decay and ingrowth with chemically and physically diverse progeny
- environmental degradation of organics

If we extend our thinking to neighboring media, we could include

- eroded sediments from the cover
- surface water runoff from the cover
- suspension of particles into the atmosphere, followed by dispersion and deposition

Many of these processes are the subject of computer models, though each model considers the process in isolation. The aforementioned hydraulic models consider water distribution and flow. Rogers and Nielson consider diffusion and decay of radon in the NRC Regulatory Guide 3.64 (NRC 1989) and other papers. Surface erosion has been often modeled simply using the Universal Soil Loss Equation and its variants, and with more complex modeling platforms such as SIBERIA (Willgoose 2005).

And yet, these processes to not operate in isolation. The deposition of radon progeny in the soil column affects the distribution of contaminants in the cover, and these would be leached by infiltrating (or exfiltrating water), for example. As burrowers churn the cover materials and the contaminants within them, plant roots are exposed to different concentrations of contaminants. As surface soils experience an influx of contaminated burrow excavations and decomposing plant parts, the contaminants entrained in wind-blown particles varies. As roots and burrows penetrate carefully-constructed layers in the cover, the layering gives way to chaos, and hydraulic properties change as a result. Changing hydraulics effects everything. The point is that all these processes are interrelated. Any attempt at understanding what is going on in the cover, much less which of these processes really matters, is limited by what is modeled in the first place.

The Need for Probabilistic Modeling

One of the challenges in such integrated modeling is that of determining what components are the most significant in any particular application. In order to assess the sensitivity of a particular result, or endpoint (e.g. the concentration of a given contaminant in some location, or the risk to a given receptor or population), one must employ sensitivity analysis (SA). In the days of slow computers, one at times had to resort to what is called one-at-a-time (OAT) SA. This was done by changing one parameter at a time and evaluating its effect on the endpoint. Not only is this approach tedious, it is severely limited, in that it depends on one's presumptive choice of which parameter to test, and is not capable of evaluating the combined effects of several parameters.

With the advent of more powerful computers and statistical techniques, we can now perform global sensitivity analyses, which evaluate the entire space of variables. These multidimensional techniques can reveal a great deal of information, leading to significant insights into the behavior of the modeled system. For example, it may reveal that at one site, the movement of water plays essentially no role in the concentration of contaminants in surface soils, but that the burrowing of ants is extremely important. In another case, we may find that the flux of radon at the ground surface is governed largely by the presence of an intact clay barrier, but that once that barrier has become permeable to gases, the flux becomes more dependent on the radon emanation factor in the wastes. Or that the dose to a receptor exposed to water from a well is driven by ingestion of tritiated water in early times, but by the ingestion of crops irrigated with water rich in technetium-99 in later time.

In addition to positively identifying which parameters matter most in a given model result, these sophisticated SA techniques can show us the range over which a particular parameter is significant. For example, changes in the soil/water partition coefficient (K_d) of plutonium may cause changes in its water well concentration only at low values of K_d, but changes at higher values show no effect. Once the SA has been completed, analysts are rewarded with a ranking of the significance of the modeling parameters, so that resources can be focused on the most important ones for the next iteration of the model.

All of these wonderful insights depend on having a probabilistic model. They also depend on having reasonable representations of the parameters that support the model. Probabilistic models are not immune to the garbage-in-garbage-out trap. In specifying the uncertainties of input parameters, one must strive to represent our state of knowledge of the problem.

Building an Integrated Model

Once we have accepted that an integrated, probabilistic model is necessary, the challenge is in its design and construction. Much can be learned from the various computer programs that are specialized for a given process, but none of them can be used directly in an integrated model. One approach that has proven successful is to build an integrated system model based on first principles, with the ability to enable and disable the various processes in order to compare them to other calculations. For example, if the integrated model includes water infiltration, other processes can be disabled, and a comparison of calculations can be made to a specialized cover infiltration model, using the same set of conditions. Similarly, one could compare radon diffusion to the analytical solution developed by Rogers and Nielson for a homogeneous material. If favorable comparisons can be made, perhaps with certain adjustments or calibrations, then this offers good support for the integrated model. Then, by enabling all processes in a simultaneous coupled solution, a model can provide a more realistic simulation of the evolution of a cover and its performance.

A fully integrated cover model is currently being developed by Neptune and Company using the GoldSim probabilistic systems analysis software. This cover model includes the following coupled processes, and can be run for any desired length of time:

- water percolation, through a solution of the Richards equation
- air phase diffusion of volatile radionuclides
- air/water partitioning using Henry's Law
- solid/water partitioning (K_d) with various porous media
- aqueous solubility limits for dissolved constituents
- radioactive decay and ingrowth
- soil excavation and collapse by burrowing animals (insects and mammals)
- contaminant uptake and translocation by plants
- changes in constructed cover layer materials and properties with bioturbation
- suspension of soil particles into the atmosphere
- surface erosion of soils and sediment transport
- subsurface flow of water diverted by cover layers

If a situation demands that the cover be assessed for extremely long time periods, such that changes in climate need to be included, for example, these effects can be included in the estimates of precipitation (and its effect on percolation), plant assemblages and productivities, and associated animal communities and activities. In short, any of these parameters or processes can be modified as time passes in the model, so that changes in environmental drivers can be captured.

This approach will offer a model of a degrading cover, with changing water flow, air diffusion, contaminant content, and transport. By studying the model results, such as fluxes of contaminants due to different processes, one can determine which of these processes is most important in a given environment. With uncertain model inputs and probabilistic execution, SA can identify the most significant input parameters, so that resources may be brought to bear on improving the model and reducing uncertainties most efficiently.

References

DOE (U.S. Department of Energy), 2006, *Addendum 2 to the Performance Assessment for the Area 5 Radioactive Waste Management Site at the Nevada Test Site, Nye County, Nevada*, DOE/NV/11718--176-ADD2, Jun 2006

Fayer, M.J., and T.L. Jones, 1990, *UNSAT-H version 2.0: Unsaturated soil water and heat flow model*, PNL-6779, Pacific Northwest Laboratory, Richland, Washington. URL: http://hydrology.pnl.gov/resources/unsath/unsath.asp

Hsieh, P.A., Wingle, William, and Healy, R.W., 2000, *VS2DI--A graphical software package for simulating fluid flow and solute or energy transport in variably saturated porous media*. U.S. Geological Survey Water-Resources Investigations Report 99-4130, 16 p. URL: http://wwwbrr.cr.usgs.gov/projects/GW_Unsat/vs2di1.2

NRC (U.S. Nuclear Regulatory Commission), 1989, *Regulatory Guide 3.64 Calculation of Radon Flux Attenuation by Earthen Uranium Mill Tailings Covers*, NRC Office of Nuclear Regulatory Research

ORNL (Oak Ridge National Laboratory), 1997, *Performance Assessment for Continuing and Future Operations at Solid Waste Storage Area 6.* ORNL/6783/R1&R2, ORNL, Oak Ridge, TN, Sep 1997

Suter, G.W. II, R. J. Luxmoore, and E. D. Smith, 1993, *Compacted Soil Barriers at Abandoned Landfill Sites are Likely to Fail in the Long Term*, Journal of Environmental Quality 22:217-226

USACE (U.S. Army Corps of Engineers), 1997, *Hydrologic Evaluation of Landfill Performance* (HELP). URL: http://www.wes.army.mil/EL/elmodels/helpinfo.html

Willgoose, G. R., 2005, *SIBERIA*, URL: http://www.telluricresearch.com/siberia-homepage.html

6.5.2   Practical Considerations for Modeling and Monitoring of Engineered Barrier Performance

Roger Seitz

Savannah River National Laboratory
Aiken, SC
Roger.Seitz@srnl.doe.gov

Design of waste disposal facilities depends on a variety of technical and non-technical considerations. In a risk-informed, performance-based view of disposal, different design options can be envisioned depending on the nature of the waste (e.g., short-lived, mobile), site-specific conditions (e.g., climate, hydrogeology), and considerations related to potential exposure scenarios (e.g., location of facility). Models are used to produce dose or risk estimates that are used as a part of the basis for decision-making, but arguably, integrated use of a combination of models and data collection to better understand system behavior is more important than the final modeling result. All of these factors need to be considered when addressing technical and policy-oriented aspects related to performance assessment modeling for disposal facilities, especially modeling approaches related to engineered barriers (e.g., NCRP 2005).

Disposal Options

A variety of different disposal options have been used and are proposed for use around the world ranging from near-surface trench disposal to disposal in deep geologic formations. The disposal options include many different concepts for engineered barriers, including: waste forms, containers, vaults/tanks, liners, covers, etc. Prescriptive designs, such as combinations of waste treatment, liners and leachate collection systems have been typically applied for hazardous waste disposal facilities. Radioactive waste disposal has tended to apply a more risk-informed approach that involves waste and site-specific design considerations (e.g., 10 CFR Part 61, DOE Order 435.1), which leads to a wide variety of disposal concepts being used in the United States.

Waste Management Policy

It is interesting to contrast some of the advantages and disadvantages of the different views of waste management. From a policy perspective, the prescriptive design approach for hazardous waste disposal facilities provides a clear path forward for development and operation of a disposal facility and minimizes the dependence on long-term, site-specific modeling. However, the prescriptive designs have tended to emphasize essentially complete isolation of the waste with active leachate management over relatively short-term time periods (e.g., 30 year time frame) and have not been as focused on longer-term performance considerations. In this approach, there can be uncertainties regarding the path forward after the initial 30 year life (e.g., how to consider degradation of engineered features, or determining when leachate collection/active maintenance can be discontinued).

The risk-informed, performance-based design approach has resulted in a variety of disposal concepts being used. This can lead to a perception of inconsistency from a policy perspective and also depends on technical reviews of modeling as well as policy decisions regarding the assessment context (time frames, receptors to be considered, etc.). However, the risk-informed approach provides the opportunity to consider short- and longer-term performance and specifically select disposal concepts based on the waste, site conditions and potential exposures and also encourages and rewards innovation. Waste disposal systems offer a diverse set of options for waste containment and control of releases to the environment (e.g., waste forms containers, vaults, liners, covers, and the site itself) and site selection can offer the opportunity to select locations that limit potential exposure pathways. Implementation of the risk-informed, performance-based approach is the focus of the presentation.

Modeling Needs and Level of Detail

The extent of modeling required for any given situation will depend on the waste, waste form, containers, engineered barriers, site conditions and exposure scenarios. For example, cover and liner performance would be expected to be much more important in the short term for uncontained wastes such as soils and debris. Likewise, cover and liner performance would not be expected to be as critical in the short term for waste that is disposed in relatively robust containers that should significantly reduce the possibility of water contacting the waste. Thus, one would expect more detailed evaluations for covers and liners that are required to perform well as opposed to cases where overall performance may not be as sensitive to cover or liner behavior. Availability of data or the ability to collect additional data will also be problem and process specific. In practice, the availability of data or relative ease of collecting data to support a more detailed model will be a significant consideration in trying to prioritize where to invest more detailed efforts.

For the reasons discussed above it is difficult to specify, *a priori*, where additional detail will be needed for any specific problem. From a practical perspective, the level of detail applied to any specific aspect of the model should have some link with the importance relative to the overall performance of the facility, data availability, and the performance objectives (i.e., the level of effort on any aspect of the problem should be somewhat commensurate with the relative importance and data availability). The philosophy of a graded and iterative approach to modeling has proven effective as a means to provide a framework to efficiently evolve a model for a specific problem in a manner that focuses the more detailed efforts for data collection and modeling on those aspects of the problem that are most significant (NAS 1990, IAEA 2000, NCRP 2005). A graded and iterative approach emphasizes the use of more simplified approaches in the initial iterations to help develop a basic understanding and begin to identify important, and unimportant, aspects of the problem to help focus efforts for subsequent iterations of the analysis. Nevertheless, it is important to take care to not miss an important factor that could get lost by using a simplification.

Practical examples of how degradation of barriers has been treated in assessments at DOE disposal facilities will be discussed in the presentation (e.g., WSRC 2008, DOE-ID 2007, DOE-ID 2010).

Monitoring Considerations

Modeling a one generally does not provide sufficient information for decision-making. Integrated approaches including a site-specific mix of monitoring, field/laboratory studies and stakeholder involvement should be used to build the case for decision-making. Monitoring is an important and visible part of an integrated approach to support operation and closure of disposal facilities and also for confirmation of effectiveness of remedial actions (e.g., IAEA 2004). Monitoring is an activity that is subject to a substantial level of public scrutiny, thus it is important that stakeholders understand the purpose and limitations of different monitoring approaches. It is critical to know how monitoring results will be interpreted (i.e., what is acceptable) and reported and the expected range of potential results before starting the monitoring. Limitations of point values, emphasis on trends rather than individual measurements, and the need to expect spatial and temporal variability (use ranges of values as the point for comparison) are important considerations when discussing monitoring programs with stakeholders. Uncertainty analyses with ranges of input parameters and outputs are better for comparison with monitoring results than deterministic simulations.

A broad spectrum of monitoring approaches can be conducted that can have different, but complementary, objectives (e.g., IAEA 2004). For the purposes of this discussion, two general types of monitoring are defined as compliance and performance monitoring, respectively. Compliance monitoring is the typical monitoring that is conducted with measurements of concentrations of contaminants in air and water being compared with specific regulatory criteria. Performance monitoring has gained interest recently and involves monitoring specific parameters of interest or indicators related to the assumptions made for risk or performance assessment models (changes in chemistry, material properties, etc.). Performance monitoring involves direct links to model assumptions and results, and thus, performance or risk assessment models are used to identify important indicators for which monitoring would be expected to provide evidence of system behavior outside of expected ranges.

Both types of monitoring serve important roles, but the distinction in objectives should be explained to stakeholders to make sure the purpose of different monitoring activities and implications of the different monitoring results is clear. For example, monitoring that detects contamination above acceptable levels in an aquifer that is used for drinking water is an immediate concern (e.g., compliance monitoring has a pass/fail context). However, while changes in chemistry in an engineered feature outside the envelope of expected values from a performance assessment may be indicative of a potential performance concern that requires more evaluation, it is not indicative of an immediate public health concern. The nature of performance monitoring is to help improve understanding of system behavior and identify potential departures from modeling assumptions, which is a different objective from compliance monitoring which is a more direct indication of a regulatory pass or fail.

Summary

There is a need for improved capabilities to model specific aspects of engineered barriers and for information gathering to improve our understanding of the performance of existing barriers and specific information regarding the durability of materials that are used in barriers. Improved models and data will help with optimization of barrier designs. However, the level of modeling detail that is needed and which aspects of the problem require more detailed models will depend on specific conditions for each facility.

When considering the level of detail to use for modeling, a graded and iterative approach is recommended to help focus more detailed modeling and data collection activities on the aspects of the system that are most significant for the decision to be made. Decisions regarding where to focus efforts will also depend on the availability of data and the ability to collect data supporting more detailed models.

Plans should include integrated approaches to build the case for decision making, including a facility-specific mix of monitoring, field/laboratory studies and stakeholder involvement as well as modeling. Models can be used to identify monitoring priorities and monitoring can be used to confirm assumptions in a performance assessment and to improve models used for performance assessments. It is important to consider how results will be reported to the public when planning innovative monitoring programs and to discuss the limitations of modeling with stakeholders to set proper expectations. Sensitivity and uncertainty analyses are useful to provide perspective regarding inputs and outputs to compare with monitoring, especially to provide ranges of values from models to compare with monitoring results.

References

DOE-ID, *Performance Assessment for the RWMC Active Low-Level Waste Disposal Facility at the Idaho National Laboratory Site*, DOE/NE-ID-11243, U.S. Department of Energy Idaho Operations Office, 2007.

DOE-ID, *Performance Assessment for the Idaho CERCLA Disposal Facility Landfill*, DOE/NE-ID-10978 Rev. 1, U.S. Department of Energy Idaho Operations Office, 2010.

IAEA, *Safety Assessment for Near Surface Disposal of Radioactive Waste*, Safety Guide, No. WS-G-1.1, IAEA Safety Standards Series, 1999.

IAEA, *Surveillance and Monitoring of Near Surface Disposal Facilities for Radioactive Waste,* Safety Series No. 35, International Atomic Energy Agency, 2004.

National Academy of Sciences/National Research Council, *Groundwater Models: Scientific and Regulatory Applications,* National Academy Press, Washington, DC, 1990.

National Council on Radiation Protection and Measurements (NCRP), *Performance Assessment of Near-Surface Disposal Facilities for Disposal of Low-Level Radioactive Waste, NCRP Report No. 152,* Bethesda, MD, 2005.

WSRC, *Performance Assessment for the F Tank Farm at the Savannah River Site*, SRS-REG-2007-00002, Rev. 0, Washington Savannah River Company, 2008.

6.5.3   Near-Term Hydrological Performance Modeling of Final Covers

Craig H. Benson, PhD, PE, DGE

Wisconsin Distinguished Professor of Geological Engineering
University of Wisconsin-Madison
www.gle.wisc.edu
chbenson@wisc.edu

Hydrologic modeling using water-balance codes that simulate cover hydrology is a key aspect of the design and performance assessment of final covers for waste containment.. Although a variety of codes have been used for cover design, only a limited number of studies have compared code predictions with direct measurements of field water-balance quantities from vegetated covers (Fayer et al. 1992, 1997; Khire et al. 1997, 1999; Roesler et al. 2002; Benson et al. 2004, 2005; Scanlon et al. 2002, 2005; Ogorzalek et al. 2008, Bohnhoff et al. 2009). Studies comparing code results with field-measured water-balance data are important because they improve confidence in codes and identify shortcomings. Code accuracy can be particularly important where a cover must be designed to meet a stringent percolation criterion, such as the 0.5 mm/yr criterion used for designing the Hanford Barrier (Fayer et al. 1992). Comparisons are also needed for covers located in different climates, constructed with different soils and/or layering, and vegetated with different plants.

Recent efforts have focused on comparing water-balance predictions from common commercial codes to field-measured water-balance data from instrumented test sections simulating full-scale covers. Measured quantities are used as input to the codes to the greatest extent practical so that ambiguity in the code evaluation is minimized. These efforts have examined different locations and climates, soil textures, the significance of vegetation and simulation of transpiration, and duration of the field data record (Scanlon et al. 2002, 2005; Ogorzalek et al. 2008; Bohnhoff et al. 2009).

This presentation describes water-balance predictions made using four codes (UNSAT-H, VADOSE/W, HYDRUS, and LEACHM) and water-balance data from final covers test sections monitored by US EPA's Alternative Cover Assessment Program (ACAP) (Albright et al. 2004). Hydraulic, meteorological, and vegetative input into the codes was based on field- and laboratory-measured properties to the greatest extent practical so that an independent assessment of the predictive capabilities of the codes could be made. Parametric simulations were also conducted to assess the sensitivity of each code to precipitation rate, saturated hydraulic conductivity, form of the hydraulic conductivity function, and the lower boundary condition.

Accuracy of the runoff prediction was found to affect the accuracy of all other water-balance quantities. Runoff was predicted more accurately when precipitation was applied uniformly throughout the day, the surface layer was assigned higher saturated hydraulic conductivity (10^{-4} cm/s), or when Brooks-Corey functions were used to describe the hydraulic properties of the cover soils. However, no definitive recommendation could be identified that will provide reasonable assurance that runoff mechanisms are properly simulated and runoff predictions are accurate. Thus, modelers should examine runoff predictions for reasonableness and consistency with other field data. Existing field data from water-balance covers suggest that runoff should not constitute more than 10% of the annual water balance.

Evapotranspiration and soil-water storage were predicted reasonably well when runoff was predicted accurately (both within ≈ 25 mm annually), general mean hydraulic properties were used as input, and the vegetation followed a consistent seasonal transpiration cycle. However, percolation was consistently under predicted even when evapotranspiration and soil-water storage were predicted reliably. Better agreement between measured and predicted percolation was obtained by increasing the saturated hydraulic conductivity of the cover soils by a factor between 5 and 10 while using general mean parameters for the soil-water characteristic curve.

References

Albright, W. H., Benson, C. H., Gee, G. W., Roesler, A. C., Abichou, T., Apiwantragoon, P., Lyles, E. F., and Rock, S. A. (2004). "Field water balance of landfill final covers." *Journal of Environmental Quality* 33(6), 1-17.

Benson, C. H., Bohnhoff, G. L., Apiwantragoon, P., Ogorzalek, A. S., Shackelford, C. D., and Albright, W. H. (2004). Comparison of model predictions and field data for an ET cover. *Tailings and Mine Waste '04*, Balkema, Leiden, The Netherlands, 137-142.

Benson, C. H., Bohnhoff, G., Ogorzalek, A. S., Shackelford, C. D., Apiwantragoon, P., and Albright, W. H. (2005). "Field data and model predictions for an alternative cover." *Waste Containment and Remediation*, A. Alshawabkeh et. al. eds., ASCE, Reston, VA, 1-12.

Bohnhoff, G., Ogorzalek, A., Benson, C., Shackelford, C., and Apiwantragoon, P. (2009), "Field data and water-balance predictions for a monolithic cover in a semiarid climate," *J. Geotech. and Geoenvironmental Eng.*, 135(3), 333-348. Fayer, M. J., and Gee, G. W. (1997). "Hydrologic model tests for landfill covers using field data." *Landfill Capping in the Semi-Arid West: Problems, Perspectives and Solutions*, Environmental Science and Research Foundation, Idaho Falls, ID, 53-68, 53-68.

Fayer, M. J., Rockhold, M. L., and Campbell, M. D. (1992). "Hydrologic modeling of protective barriers: Comparison of field data and simulation results." *Soil Society of America Journal*, 56, 690-700.

Khire, M., Benson, C. H., and Bosscher, P. J (1997). "Water balance modeling of earthen final covers." *Journal of Geotechnical and Geoenvironmental Engineering*, 123(8), 744-754.

Khire, M., Benson, C., and Bosscher, P. (1999). "Field data from a capillary barrier and model predictions with UNSAT-H." *Journal of Geotechnical and Geoenvironmental Engineering*, 125(6), 518-527.

Ogorzalek, A., Bohnhoff, G., Shackelford, C., Benson, C., and Apiwantragoon, P. (2008). "Comparison of field data and water-balance predictions for a capillary barrier cover." *Journal of Geotechnical and Geoenvironmental Engineering*, 134(4), 470-486.

Roesler, A. C., Benson, C. H., and Albright, W. H. (2002). *Field Hydrology and Model Predictions for Final Covers in the Alternative Assessment Program. Geo Engineering Report No. 02-08*, University of Wisconsin-Madison.

Scanlon, B. R., Christman, M., Reedy, R. C., Porro, I., Šimůnek, J., and Flerchinger, G. N. (2002). "Intercode comparisons for simulating water balance of surficial sediments in semiarid regions." *Water Resources Research*, 38(12), 59-1 - 59-16.

Scanlon, B. R., Reedy, R. C., Keese, K. E. and Dwyer, S. F. (2005). "Evaluation of evapotranspirative covers for waste containment in arid and semiarid regions of the Southwestern USA." *Vadose Zone Journal*, 4, 55-71.

6.5.4   Simulating the Long-term Performance of Engineered Barriers in Arid Environments with the STOMP Sparse Vegetation Evapotranspiration Model

A. L. Ward

Pacific Northwest National Laboratory
P.O. Box 999 MSIN K9-33, Richland, WA 99352

It is anticipated that some 200 surface barriers, covering nearly 1000 acres, will be built at Hanford to reduce the local flux of meteoric water into subsurface waste zones (Ward et al. 2005). Design and licensing of these barriers require a demonstrated understanding of the multidimensional, nonisothermal geohydrologic and coupled ground surface to atmosphere water mass and energy transport process that control percolation. However, owing to the complex interactions between the atmosphere and barrier surface and between the barrier surface and surface, barrier design remains one of the biggest contributors to the uncertainty in long-term performance. At waste sites in arid and semi-arid environments, barrier design must consider freezing conditions, surface runoff, subsurface capillary breaks, lateral drainage, side-slope performance, and evapotranspiration from sparse canopies as well as the impacts of spatial and temporal changes in physical and hydraulic properties that might influence long-term (> 1000 yrs) performance (White and Ward, 2005). The available design tools that can meet these requirements are quite limited.

The barrier design tool describe in this paper is based on Water-Air-Energy Operational Mode of the Subsurface Transport Over Multiple Phases (STOMP) simulator (White and Oostrom, 2000). A soil-vegetation-atmosphere-transfer scheme based on a sparse vegetation evapotranspiration model was used to extend the Water-Air-Energy Operational Mode of STOMP for predicting performance of engineered barriers in arid and semi-arid environments (Ward et al, 2005; White and Ward, 2005). A major advancement relates to the simulation of evapotranspiration. The pre-computer era equations of Penman (1948) and Monteith (1965), developed for full canopies free of water stress, and used in typical barrier design tools have been replaced. The Shuttleworth-Wallace (1985) evapotranspiration model, developed for sparse canopies, is extended to account for multiple plant species and to eliminate the need for a surface resistance to account for reduced unsaturated water vapor conditions at the soil surface. Another significant improvement relates to the estimation of the evaporative depth, a parameter in loosely coupled models that be changed at will to dictate barrier performance. In STOMP mass and energy transfer are tightly coupled and the evaporative depth, which is essentially the lower boundary of the dry soil layer formed during the drying process, is determined implicitly. In this unique approach, the evaporative depth is determined by balancing the water and energy fluxes as the SVAT process proceeds.

With this extension, the model calculates water mass, air mass, and thermal energy across a boundary surface and root-water transport between the subsurface and atmosphere. The model solves sets of nonlinear conservation equations for water mass, air mass and thermal energy at the ground surface, plant leaves, and canopy. The conservation equations mathematically describe the transport of water, air, and thermal energy across the ground surface, either directly or through plants. Nonlinearities in the solved governing equations are resolved via Newton-Raphson iteration. The subsurface equations are coupled to the surface evapotranspiration equations as a boundary condition, whose effects impact deeper subsurface nodes through plant roots. The sets of solved governing conservation equations differ, depending on whether the ground surface is bare or vegetated and whether the temperatures of different plant species are to be distinguished. The bare-surface system of equations comprises steady-flow conservation equations for water mass, air mass, and thermal energy at the ground surface. The air-mass equation is implicit, yielding a system of two equations and two unknowns to be resolved at each boundary surface.

The single-plant-temperature system of equations comprises steady-flow conservation equations for water mass at the ground surface and mean canopy height (canopy) and thermal energy at the ground surface, plant leaves, and canopy. An air-mass conservation equation at the ground surface and water-mass conservation equation at the plant leaves are implicit, yielding a system of five equations and five unknowns to be resolved at each boundary surface. There is an option to account for rainfall interception and condensation accumulation in which case the water mass conservation equation at the plant leaves changes from steady-flow to transient, including plant-leaf storage of water. Principal design goals of the STOMP barrier simulation tool include broad applicability, verified algorithms, quality assurance controls, and simulations validated against laboratory and field-scale experiments. The calibration of key input parameters, including hydrologic, thermal, in heterogeneous anisotropic systems and is afforded via inverse capability built into the model (Zhang et al., 2002). Forward and inverse simulation of complex field-scale systems is afforded though the use of new scaling techniques (Zhang et al., 2004; Ward et al. 2006) and implementation for scalable execution on multiple processor (i.e., parallel) computers (Ward et al. 2008).

For verification of the model, a series of eleven problems related to infiltration, redistribution, evaporation and transpiration in homogeneous and layered soils have been simulated (Ward et al. 2005). A comparison of simulation results with published analytical and numerical solutions and experimental results demonstrates that STOMP is able to describe the dynamics of mass and energy transport over a range of meteorological and soil conditions with or without plants present. In addition, water balance calculations have been performed for a grass-covered site in Hanford's 300 Area and a bare-surface monofill barrier in Idaho (Scanlon et al., 2002). Best-estimate values for the hydraulic and thermal parameters for soil and geotechnical materials, and ecological parameters for simulating plant response have been developed for Hanford (Ward, 2007). Use of field-scale hydraulic properties determined for the grass site by automatic calibration resulted in very good agreement between observed and predicted water balance. In an inter-code comparison with seven public domain codes (UNSAT-H, HYDRUS-1D, SHAW, SWIM, VS2DTI, SOILCOVER, and HELP) used to simulate barrier performance, the STOMP simulator was better able to match field observations. Use this tool to guide barrier design at arid and semi-arid sites will result in optimized designs that reduce construction costs, reduce environmental impacts at borrow sites, and minimize post-closure care and monitoring needs while meeting regulatory requirements.

References

Monteith JL. 1965. "Evaporation and environment." p. 205-234. In: *State and Movement of Water in Living Organisms.* 19th Symposium of the Society for Experimental Biology. Cambridge University 111:839-855. Cambridge University Press, Cambridge.

Penman HL. 1948. "Natural evaporation from open water, bare soil, and grass." *Proc. Royal Soc. London* A193:120-146.

Scanlon BR, M Christmans, RC Reedy, I Porro, and J Simunek, and GN Flerchinger. 2002. "Intercoce comparisons for simulating water balance of surficial sediments in semiarid regions." *Water Resour. Res.* 38(2):1323.

Ward AL. 2007. *Geotechnical, Hydrogeologic, and Vegetation Data Package for 200-UW-1 Waste Site Engineered Surface Barrier Design*, PNNL-17134, Pacific Northwest National Laboratory, Richland WA.

Ward AL, M Oostrom, and DH Bacon. 2008. *Experimental and Numerical Investigations of Soil Desiccation for Vadose Zone Remediation: Report for Fiscal Year 2007*. PNNL-12030, Pacific Northwest National Laboratory, Richland, WA.

Ward, A.L. and Z.F. Zhang. 2007. Effective Hydraulic Properties Determined from Transient Unsaturated Flow in Anisotropic Soils. Vadose Zone Journal. 6:913–924.

Ward AL, ZF Zhang, and GW Gee. 2006. Upscaling unsaturated hydraulic parameters for flow through heterogeneous anisotropic sediments. Advances in Water Resources. 29:268-280

White MD, and AL Ward. 2005. "Numerical simulation of surface barriers for shrub-steppe ecoregions.' *Hydrology Days* 25:224-236.

White MD, and M Oostrom. 2000. *STOMP Subsurface Transport Over Multiple Phases,* Version 2.0, Theory Guide. PNNL-12030, Pacific Northwest National Laboratory, Richland, WA.

Zhang ZF, AL Ward, and GW Gee. 2002. *Estimating Field-Scale Hydraulic Parameters of Heterogeneous Soils Using A Combination of Parameter Scaling and Inverse Methods.* PNNL-14109, Pacific Northwest National Laboratory, Richland, WA.

Zhang ZF, AL Ward, and GW Gee. 2004. "A parameter scaling concept for estimating field-scale hydraulic functions of layered soils." *J. Hydraulic Res.* 42:93-103.

6.5.5 Effects of Plant Succession on the Functioning of Engineered Covers and Modeling of Long-Term Successional Impacts using the EDYS Ecological Simulation Model

Terry McLendon

KS2 Ecological Services, Fort Collins, Colorado

Three major functions of engineered covers placed over waste materials are to 1) provide a substrate for the establishment and maintenance of an adequate plant community to stabilize the surface, 2) minimize infiltration of water into the waste material, and 3) minimize any translocation of waste materials to the surface via plant root uptake. Successful accomplishment of these three functions requires that the vegetation on the site, both initial and that which develops over time, have the proper combination of both above- and belowground components.

Required aboveground components of the vegetation are relatively simple. The vegetation must be adequate to stabilize the soil surface against wind and water erosion and must provide a sufficient level of productivity to remove the desired amount of soil water through evapotranspiration (ET). Other factors (e.g., adaptability to grazing, both livestock and wildlife; minimize impacts from burrowing animals; ability to recover from fire and other surface disturbances; esthetic considerations) become important in many cases, but the primary functions are provision of adequate erosion control and ET.

Belowground components of the vegetation are more complex in the successful design and implementation of covers. Roots are adapted to exploiting belowground resources, especially water, under an almost infinite variety of substrate conditions. Roots grow into the smallest of cracks and fissures and expand these openings, creating large openings that are then filled with additional roots and debris, and can be further exploited by burrowing animals. These openings become channels for downward movement of soil water, which contributes further to opening of the substrate material through growth of more roots, colonization by animals, and by freezing and thawing in temperate and northern climates. The potential extent and magnitude of these root-induced changes depend on a number of factors, including: type and species of the plants, productivity of the aboveground plant community, land-use on the surface of the cover, availability of soil water in the various soil layers, thickness and texture of the overlying material, and density and thickness of the barrier.

Of primary importance are the lifeform and species of plants and the characteristics of the materials overlying the barrier. Species rooting characteristics are both genetically and environmentally determined. Each species has a maximum potential rooting depth that is genetically set. Environmental conditions, such as soil texture, bulk density, depth to saturated soil, often restrict this genetic limit. Grasses typically have maximum rooting depths on the order of 1-2.5 m, but some species can root to over 6 m. In general, shrubs have deeper maximum rooting depths, generally on the order of 2-6 m, with some shrubs rooting to 9 m. Trees have the maximum rooting depths, with numerous species rooting to 10-20 m, and the maximum reported for mesquite is 53 m. Maximum potential rooting depth tends to be approached on sandy or loamy soils with abundant soil moisture at depth. Clay soils, saturated layers, and dense layers such as claypans or indurated caliche, often restrict rooting depth to less than the maximum potential depth.

Roots are not evenly distributed throughout the rooting depth. Most root biomass in concentrated in the upper portions of the rooting zone. For many species, half or more of their total root biomass is concentrated in the upper 20% of the root zone. These upper portions of the soil profile tend to contain the largest proportion of water and nutrients, especially nitrogen, in the soil profile and the concentration of roots in these upper layers allows the plants to maximize uptake. Deeper roots are important for stability and for securing deep sources of nutrients and soil water, especially during dry periods. It is generally these deeper roots that pose the greatest threat to subsurface cover barriers.

Plant succession is the ecological process of directional vegetation change over time, usually beginning with relatively-short lived herbaceous plants and culminating in plant communities dominated by long-lived, generally woody, species. Succession occurs on all sites. The rate of succession can be relatively rapid, especially in regions of higher rainfall, or it can be very slow, as in some desert and arctic regions, but this process of vegetation change is constantly taking place. In the absence of continual maintenance programs (e.g., brush control, mowing, tree removal), the vegetation on the cover of a waste site will change over time.

Two common aspects of succession are 1) an increase in vegetation structure and 2) an increase in the relative amounts of woody plants. Both of these aspects have profound implications to the function of engineered barriers. Vegetation structure refers to the shape of the vegetation, e.g., height, coverage, and stratification. Structure increases as succession proceeds, both above- and belowground. Aboveground, the height of the vegetation increases (e.g., grasses are replaced by shrubs and then perhaps by trees), coverage of the soil surface increases, and layering (strata) of vegetation occurs, with different species occupying different vertical layers. Similar processes occur belowground. Root systems become deeper as shallow-rooted species are replaced by deeper-rooted species, root biomass increases in lower soil depths as the number and types of species increase, and the density of the root system increases in the various layers. These deeper and more dense root systems place more stress on the underlying barrier and potentially can exploit more of the cracks and fissures in the barrier material.

As the vegetation shifts from dominance by herbaceous plants (e.g., grasses), which have relatively shallow root systems but with very dense root mass in the upper profile, to dominance by woody species (e.g., shrubs and trees), which have deeper roots systems with proportionately more roots in deeper layers, the hydrological dynamics of the system change. Early successional plant communities tend to extract most of the water they transpire from the upper soil profile. Late successional communities have greater ability to extract water from depth. This can be both a positive and a negative in the functional efficiency of covers.

Because of successional changes in the vegetation, the plant-soil-water characteristics of a cover are likely to become very different over time. Conditions 50-100 years after construction are not likely to be similar to those soon after construction was completed. In some ways, conditions will be more favorable, e.g., ET will likely be higher thus reducing the amount of deep infiltration and stability of the vegetation may be greater. In other ways, conditions will be less favorable, e.g., deeper root systems will place much greater stress on the buried barrier. Because succession is a process that is near-universal ecologically, these changes should be accounted for in the design of cover systems. The only alternative is perpetual maintenance.

One simple method of engineering to account for succession is to place a thick layer of inexpensive substrate material between the cover barrier and the upper topsoil. This material can consist of almost any locally available fill material: sand, gravel, rock, low-grade clay, non-hazardous non-organic material. The purpose of this layer is to provide sufficient depth between the surface and the barrier to assure that only minimal amounts of roots will ever reach the top of the barrier. The actual depth will depend on the potential late-successional vegetation at that site.

Simulation modeling provides a useful tool in designing covers to take into account succession and other types of vegetation change. One example that has been used for developing and evaluating cover designs and for watershed management is the Ecological Dynamics Simulation (EDYS) model. EDYS is a spatially-explicit mechanistic model that simulates vegetation dynamics (above- and belowground), hydrology (surface erosion, ET, infiltration, percolation, subsurface drainage), and the effects of natural and anthropogenic impacts (e.g., fire, grazing, climate change) on the landscape at spatial scales ranging from less than 1-m^2 to watershed-level (100,000$^+$ hectares) and at time scales ranging from days to centuries. EDYS has been used as a tool in the development of cover designs from Montana and Washington to Arizona and in Indonesia. It is currently being used to evaluate the effects of land-use scenarios on vegetation change and water-use on watersheds in Texas for the US Army Corps of Engineers and in California of the Los Angeles Department of Water and Power. It has been previously applied to watershed projects and management of military lands in nine states and in Australia.

Figure 1 provides an illustration of the use of EDYS to evaluate changes in root development over time. Simulations are from the evaluation of the cover design for the Mineral Hill TVX mine site in Montana (McLendon et al. 2002, Amerikanuak 2006). Year 1 values are simulated root biomass by soil layer (four layers of topsoil and four layers of sand over tailings) at the end of the first growing season following revegetation of the site. The plant community was a seeded grassland with Douglas fir seedlings. Year 10 values are the simulated root biomass values at the end of 10 years of succession, as the trees begin to dominate the site. By Year 10, roots have penetrated into the underlying tailings. In this design, there was no engineered barrier between the lowest sand layer and the upper portion of the tailings.

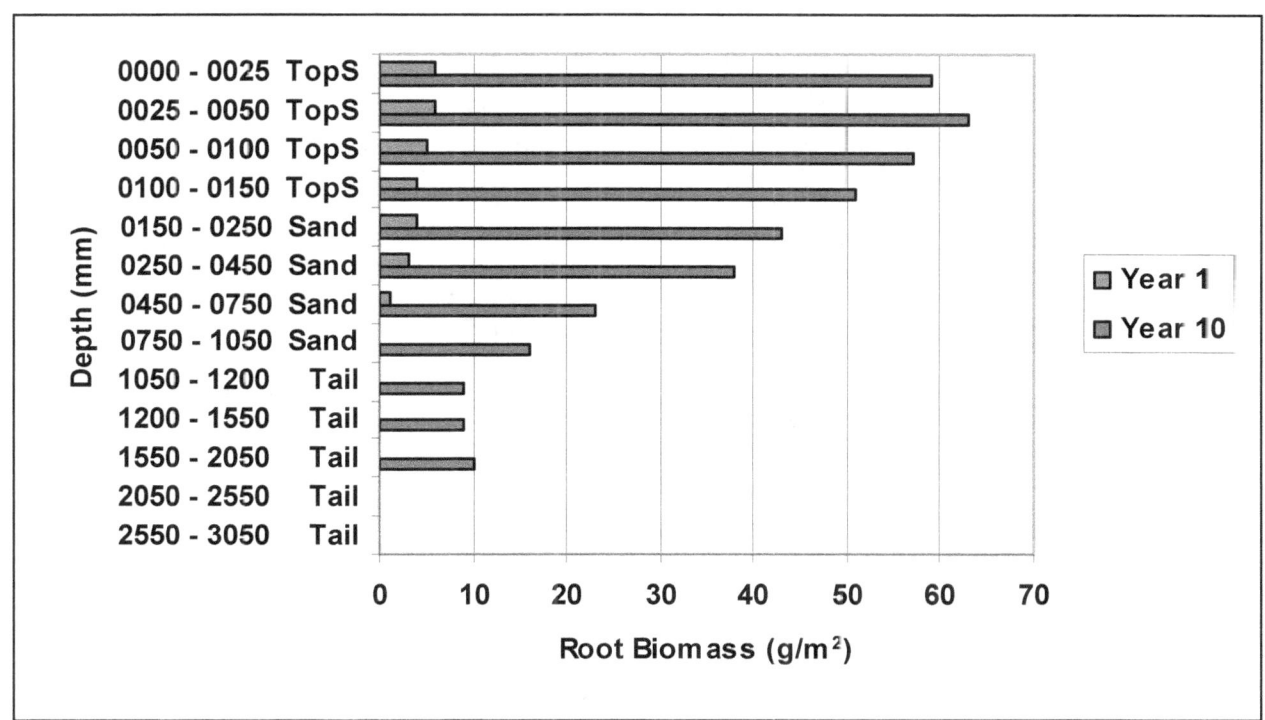

Figure 1. EDYS simulations of root development on the water-balance cover over tailings at the Mineral Hill TVX mine site, Montana.

References

Amerikanuak, Inc. 2006. Proposal for the TVX Mineral Hill Mine consolidated closure plan modifications near Gardiner, Montana. Report submitted to the Montana Department of Environmental Quality. Submitted by TVX Mineral Hill, Inc. Toronto, Canada. 139 p.

Childress, W. Michael, David L. Price, Cade L. Coldren, and Terry McLendon. 1999. A functional description of the Ecological Dynamics Simulation (EDYS) model, with applications for Army and other Federal land managers. US Army Corps of Engineers CERL Technical Report 99/55. 68 p.

Childress, W. Michael, Cade L. Coldren, and Terry McLendon. 2002. Applying a complex, general ecosystem model (EDYS) in large-scale land management. Ecological Modelling 153:97-108. linkinghub.elsevier.com/retrieve/pii/S030438000100504X

Johnson, Billy E. and Cade L. Coldren. 2006. Linkage of a physically based distributed watershed model and a dynamic plant growth model. ERDC/EL TR-06-17. US Army Corps of Engineers. Engineer Research and Development Center. Vicksburg, Mississippi. 95 p.

Mata-Gonzalez, Ricardo, Rachael G. Hunter, Cade L. Coldren, Terry McLendon, and Mark W. Paschke. 2007. Modeling plant growth dynamics in sagebrush steppe communities affected by fire. Journal of Arid Environments 69:144-157.
www.elsevier.com/locate/jaridenv

McLendon, Terry, Jeffrey Coleman, Thomas A. Shepherd, and Robert E. Nelson. 1997. The inclusion of biointrusion considerations in the design of the reclamation cover for the DMC tailings impoundments. Tailings and Mine Waste Conference 97 Proceedings. Balkema. Rotterdam. pp 267-281.

McLendon, Terry, W. Michael Childress, Cade Coldren, and David L. Price. 2001. EDYS experimental and validation results for grassland communities. US Army Corps of Engineers ERDC/CERL TR-01-54. 88 p. handle.dtic.mil/100.2/ADA396255

McLendon, Terry, W. Michael Childress, Cade L. Coldren, Rick Frechette, and Frank Bergstrom. 2002. Evaluation of alternative cover designs for a water-balance cover over tailings at the Mineral Hill Mine, Montana, using the EDYS model. Tailings and Mine Waste 02. Proceedings of the Ninth International Conference on Tailings and Mine Waste. Balkema, Rotterdam. pp 505-518.

McLendon, Terry, Cade L. Coldren, and David L. Price. 2009. Comparison of results from the EDYS and EDYS-L ecological simulation models as applied to vegetation and hydrological dynamics on the Honey Creek Watershed, Texas. Technical Report ERDC TN-SWWRP-09-7. US Army Corps of Engineers. 19 p. http://el.erdc.usace.army.mil/elpubs/pdf/swwrp-09-7.pdf

Price, David L., Terry McLendon, and Cade L. Coldren. 2004. Application of an ecological model for the Cibolo Creek Watershed. Water Quality Technical Notes Collection. ERDC WQTN-CS-04. US Army Engineer Research and Development Center. Vicksburg, Mississippi. http://www.dtic.mil/cgi-bin/GetTRDoc?AD=ADA423809&Location=U2&doc=GetTRDoc.pdf

United States Air Force Academy. 2000. Environmental Assessment Analysis of Jack's Valley Operations. Environmental Engineering Flight. USAF Academy. Colorado Springs, Colorado.

6.5.6 Application of Thermal Remote Sensing for Multi-Scale Monitoring of Evapotranspiration

William P. Kustas and Martha C. Anderson

USDA-ARS, Hydrology and Remote Sensing Laboratory, Beltsville, MD 20705

Thermal infrared (TIR) remote sensing of land-surface temperature (LST) provides valuable information about the sub-surface moisture status affecting evapotranspiration (ET) (Anderson and Kustas, 2008). There exists a plethora of techniques exploiting LST and vegetation amount (e.g., as quantified by vegetation indices such as the Normalized Difference Vegetation Index; NDVI), many of which are empirical or are based loosely on conceptual/physical processes simulated by more complex soil-vegetation-atmosphere-transfer (SVAT) schemes (Kalma, 2008). However, often such remote sensing-based methods provide ambiguous results since other factors, related to meteorological (radiation, advection, air temperature) and landscape conditions (land use, roughness, fractional cover) are affecting plant water use as well as the relationship between soil and canopy component temperatures on the composite LST (Kustas and Anderson, 2009). A more physically-based interpretation of LST and NDVI and their relationship to surface and sub-surface moisture conditions can be obtained with a simplified SVAT driven by TIR remote sensing. This approach, the Atmosphere-Land Exchange Inverse (ALEXI) model (Anderson et al., 2007a), couples a two-source (soil + canopy) land-surface model with an atmospheric boundary layer model in time-differencing mode to routinely and robustly map soil and vegetation fluxes across the U.S. continent at 5-10 km resolution using thermal band imagery from the Geostationary Operational Environmental Satellites (GOES). The soil moisture condition is quantified in terms of the reduction of evapotranspiration (ET) from the potential rate (PET) expected under non-moisture limiting conditions. A derived Evaporative Stress Index (ESI), given by anomalies in 1-ET/PET, shows good correspondence with standard drought metrics and with patterns of antecedent precipitation, but at significantly higher spatial resolution due to limited reliance on ground observations (Anderson et al., 2007bc). Higher resolution ET assessments can be generated through spatial disaggregation scheme called DisALEXI (Norman et al., 2003; Anderson et al, 2007a) using TIR data from polar orbiting instruments such as Landsat (60-120m) and MODIS (1km). This allows monitoring of ET at the subwatershed and field scales, and moreover permits validation of the ET product using tower-based flux observations.

The ALEXI algorithm is diagnostic and does not require precipitation or soil texture information, unlike most other physically-based ET models. This is a significant advantage for monitoring many areas where such information is not locally available, particularly precipitation which must be extrapolated from a sparse observation network. It has also prompted the development of assimilation schemes using ALEXI model output of ET and a soil moisture proxy to constrain water balance-SVAT model predictions (Crow et al., 2008; Hain et al., 2009).

References

Anderson, M.C., Kustas, W.P., 2008. Thermal remote sensing of drought and evapotranspiration.Eos, Trans. AGU 89, 233–234.

Anderson, M.C., Kustas, W.P., Norman, J.M., 2007a. Upscaling flux observations from local to continental scales using thermal remote sensing. Agron. J. 99, 240–254.

Anderson, M.C., Norman, J.M., Mecikalski, J.R., Otkin, J.P., Kustas, W.P., 2007b. A climatological study of evapotranspiration and moisture stress across the continental U.S. based on thermal remote sensing: I. Model formulation. J. Geophys. Res. 112, D10117, doi:10.1029/2006JD007506.

Anderson, M.C., Norman, J.M., Mecikalski, J.R., Otkin, J.P., Kustas, W.P., 2007c. A climatological study of evapotranspiration and moisture stress across the continental U.S. based on thermal remote sensing: II. Surface moisture climatology. J. Geophys. Res. 112, D11112, doi:10.1029/2006JD007507.

Crow, W.T., Kustas, W.P., Prueger, J.H., 2008. Monitoring root-zone soil moisture through the assimilation of a thermal remote sensing-based soil moisture proxy into a water balance model. Remote Sens. Environ. 112, 1268–1281

Hain, C.R., Mecikalski, J.R., Anderson, M.C., 2009. Retrieval of an available waterbased soil moisture proxy from thermal infrared remote sensing. Part I: methodology and validation. J Hydrometeol. 10, 662-682.

Kalma, J.D., McVicar, T.R., McCabe,M.F., 2008. Estimating land surface evaporation: A review of methods using remotely sensing surface temperature data. Surv. Geophys., doi:10.1007/s10712-008-9037-z.

Kustas, W.P., Anderson, M.A., 2009. Advances in thermal infrared remote sensing for land surface modeling. Agric.Forest Meteor. 149, 2071–2081,

Norman, J.M., Anderson, M.C., Kustas, W.P., French, A.N., Mecikalski, J.R., Torn, R.D., Diak, G.R., Schmugge, T.J., Tanner, B.C.W., 2003. Remote sensing of surface energy fluxes at 10^{1}-m pixel resolutions. Water Resour. Res. 39, doi:10.1029/2002WR001775.

SESSION 5:

EXPERIENCE WITH MODEL SUPPORT AND MULTIPLE LINES OF EVIDENCE TO GAIN CONFIDENCE IN LONG-TERM PERFORMANCE

Session Chairs:
Hans Arlt, NRC/FSME and
George Alexander, NRC/FSME

Technical Reporter:
Brooke Traynham, NRC/FSME

7.1   Background and Motivation

Engineered surface barriers can be significant barriers. They may help provide reasonable assurance that statutory requirements will be met. Numerical models are increasingly being used to estimate and project future risks from a variety of facilities associated with the disposal of nuclear materials including disposal sites using engineered barriers. It is desirable to validate model results by collecting new post-modeling data to determine if the model predictions were correct. This can be done for model runs with relatively short-time periods. Numerical models for radioactive waste disposal cannot be validated in a traditional sense if the performance period extends over hundreds or thousands of years. Since performance assessment results are only as good as the input and support provided for the models, model support and confidence building are important in demonstrating that reliance on model results with inherent uncertainties will not endanger public health and safety. Together, numerical modeling and model support can be used to build confidence in engineered barriers that take credit for long-term performance.

Model support is the use of multiple lines of evidence to provide confidence that the projected future performance of the system will reasonably approximate actual future performance. Model support can use multiple types and sources of information and may include: site specific tests; information on previous experience with similar systems; process models of barrier component performance; natural and anthropogenic analogs; independent peer review (expert elicitation); or plans to develop additional model support for engineered surface barrier system performance. Adequate technical bases are needed so that the modeling is appropriately accounting for uncertainties or is sufficiently conservative based on the model support. Technical bases and model support can help provide the knowledge needed to understand interrelationships between processes, e.g., between the extent of cover erosion and the infiltration rates into the cover, and provide reasonable assurance the coupled processes are adequately understood.

This session included discussions on the different types of model support strategies and multiple lines of evidence and how they can be used so as to build confidence in the model results. Topics discussed included: field evidence and laboratory tests to build confidence in performance; ACAP exhumation and process audits to identify failure modes; lessons learned from uranium recovery experiences and monitoring programs; model support commensurate with the risk significance; plant succession and soil development affecting long-term performance; landform stability as analogs to engineered barriers; attributes and evolution of stable landforms; time periods for evaluation; and the applicability of performance confirmation programs. Organizers and participants were interested in knowing what information sources or "lines-of-evidence" are needed to have confidence that an engineered surface cover or bottom liner will perform as predicted for 100 years, or for 100's to 1000's of years as ecologic settings and climates change?

7.2   Highlights of Session Presentations

David W. Esh, NRC staff, provided an overview of model support, as an important element in the application of engineered barriers to limit risk. Model support principles ensure the likelihood of making a good decision based on performance projections. Best practices include: multiple lines of evidence, direct observations, expert elicitation, accelerated experiments, natural analogs, and support for the full range of expected conditions. Natural analogs provide some data for very long-term assessments; however, they can have a confirmation bias (other analogs may have weathered away), and past exposure conditions can be unknown and highly dynamic. It is important to remember that the complexity of the model should be no greater than the model information supporting it.

Abraham Van Luik, U.S. DOE, discussed activities that support the scientific credibility of radioactive-waste, system-performance models. His presentation emphasized that model support is confidence building, and that model support activities should reflect the potential disposal-facility risks. Model validation includes corroborating model information with as many lines of evidence, as possible, in both the present and post-development confidence building activities. This includes single realization analyses to investigate the interactions, and the cause-effect relationships between model components, in addition to assisting in error checking and model verification. Uncertainty analyses should be subdivided into epistemic and aleatory uncertainties.

The features, events, and processes (FEPs) evaluation process provides a structured approach to identify relevant and non-redundant elements to be analyzed. Those FEPs that pass through a screening test (i.e., those credible and likely to occur) will be included in the model. Performance-margin analysis includes less conservative alternative conceptual models. Compounded conservativisms may serve to dilute risk ultimately by adding non-conservativisms to the system. Both natural and anthropogenic analogs can support model development (e.g., water seepage into a subsurface void as an analog for water seeping into a man-made tunnel). Independent technical review process provides peer review recommendations. Performance confirmation monitoring and testing are conducted to evaluate the adequacy of the information used to demonstrate safety by confirming that the system is operating as originally intended. Competing objectives and goals require tradeoffs which should be explicitly identified and acknowledged.

George Koerner, Geosynthetic Institute (GSI) presented an overview of geomembrane (GM) performance. Factors affecting lifetime performance included ultraviolet light, radiation, oxidation, elevated temperatures, and various types of stress. The lifetime of the GM will be dependent upon the extent of exposure to these degradation processes. There are three stages of degradation: (1) antioxidant depletion, (2) induction time, and (3) half-life of property. A reaction equation that includes these parameters will allow for calculation of overall performance lifetime. Buried lifetime prediction needs to be well researched. There are also lifetime equations for exposed GM's. Different types of GM's have different lifetime performance potentials. Recommendations made at the workshop included a comprehensive system approach. This approach included: a double composite cover that would performance best for the required periods of performance. Due to its expected performance, quality assurance (QA) is needed during construction to include a performance test facility before the cover is to perform, and rigorous maintenance (especially for exposed GMs). Since geomembranes are fragile materials, strict QA/QC is critical to long-term performance.

Jody Waugh, S.M. Stoller LLC, presented an overview of the role of natural analogs in the design and long-term performance evaluation of earthen covers for uranium mill tailings, and the ways they may be of value as a line of evidence for model support. In the long term, the ecology of a cover will change in ways that cannot be accurately predicted by models alone. Ecosystem engineering paradigm is a useful way to view the cover system since earthen covers are engineered ecosystems that will be greatly influenced by the surrounding ecosystem. Analogs provide tangible evidence and clues for understanding these changes, and provide evidence for: (1) monitoring precursors to change, (2) designing field experiments, and (3) developing and screening scenarios for performance modeling. Several natural analog examples were presented: For the Monticello site evapotranspiration (ET) cover, the major issue of long-term shift and variability in climate was evaluated using analogs of past climate (e.g., tree rings, packrat middens, lake pollen, ice cores, and archeology) coupled with climate change models. Paleoclimate data provides more local data than regional climatic-change models (due to finer-scale data). Shifts in plant abundance provide information on how precipitation and temperature conditions may have changed over time. Fire chronosequences provides evidence of recovery after a fire disturbance. Side slope stability can be examined using an "ancient-hill" analog or glacial debris flow (to examine long-term erosion protection). Pedogenic carbonates provide analogs of a capillary barrier on the order of 10,000 years by examining pedogenic tracer of long-term water movement. It was recommended that patchworks of vegetation should be examined as microenvironments within the total cover system.

Todd Caldwell, Desert Research Institute/UNV, discussed arid soil evolution and pedologic development applicable to engineered barrier designs. He mentioned that drainage is almost entirely absent during arid soil evolution and pedogenic development where soils lock everything in place. Soil development will be a function of climate, parent material, topography, biology, and time. Pedogenic features will affect hydraulic processes and plant communities at various scales, both spatially and temporally. The modern landscape provides a record of soil evolution pathways that can support long-term assessments of covers. Hydraulic properties of soil are a function of both biotic and abiotic processes such as roots growth, faunal burrowing, nutrient cycling, and resource translocation. Soils undergo a development of macropores with time, transitioning from micropore dominated soils. There is a predictable decrease in the permeability of the upper horizon when starting with lose wind-blown sediments, resulting in more water being retained at the surface. Young soils have no development, implying sandy texture with high infiltration while old deposits are clay rich. Incipient soil formation involves short-term soil development processes relevant to cover evolution. Pedogenesis for a compacted clay, or engineered silt loam will likely evolve much faster. Vegetation distribution and size is largely a function of soil-grain size and composition. Feedback between soil and vegetation will influence: canopy height/volume; shrub abundance; rooting depth; and lateral spread/plant type.

7.3   Recommendations/Insights from Panel Discussion

Panel members and workshop participants discussed the following recommendations during the panel discussions:

-   Analysts must make decisions with imperfect information and be ready to adapt to new information/observations/model support. Model support should be risk-informed and needs to evolve with different stages of the project. It should be specific for each site characteristic.

-   Develop and implement strategies to obtain and evaluate information needed to support both short- and long-term modeling results.

-   Level of engineering required should be aligned with the level of risk; the costs of monitoring; the cost of remediation and cleanup; and the total system performance. If the cover is critically important, then there should be a redundant system. A discussion of the covers should not be made in isolation.

-   The cover normally incorporates a defense-in-depth approach. For mill tailing sites, multiple layers can include layered tailings with the less radioactive material overlying the more radioactive material, a soil/frost protection layer, in addition to a radon barrier, and siltation of the rock covers. However, most panelists stressed the need to measure radon flux or concentration at the surface.

- Engineering design should be reassessed. For example, water movement may be minimized by using a modular design, as opposed to a monolithic design, resulting in ground-water recharge between modules to increase dispersive mixing and dilution.

- The important properties of clay barriers quickly degrade when exposed to the surface environment (i.e., wet-dry and freeze-thaw cycles). By locating the clay barriers sufficiently below an overlying cover (e.g., soil) unit, certain degradation processes may be minimized. More information is needed on the relationship between the depth of the soil (clay) barrier and the thickness of the overlying cover material that may affect the rate of degradation or property changes.

- Set up multiple experiments and physical tests for future generations to study the burial of materials and exhumation after 50-100 years.

- Concentration profiles are not useful for fluids. However they may be helpful for the soil column influencing the transport processes. Hybrid monitoring and modeling approaches could be useful. This would include: (1) conducting basic surveys first to understand the system; (2) parameterizing the system by creating pedotransfer functions in order to spatially distribute samples; and (3) distributing the parameters over the range of transport times and conduct tracer experiments. Evaluation of transport times may require the analyst to go back to step one due to abnormalities, sampling locations, or instrumentations. Model abstraction must be revisited. Ideally, a large range of abstractions enable evaluation of many different system attributes.

- Account for episodic events, as well as for average rates.

- The model is no better than the assumptions describing it. Common mistakes are too frequently made in cover design during the performance assessment process. An independent review can minimize potential errors. Assumptions should not be too restrictive to negate the total system performance.

- Model support should be risk-informed, and needs to evolve with different stages of the project. It should be specific to each site's characteristics. For example, as new information becomes available, it should be used in the actual model development and construction. A need exists for clear feedback between the assumptions that the modeler makes, and the actual construction data and monitoring years later.

- Develop a "Catalogue of Natural Analogs" for different climatic and environmental settings throughout the U.S.

- A catalog of analogs could provide insights into ecology, climate, and erosive forces for long-time periods, and could provide bounding scenarios of reasonable future states. For example, pedogenic carbonates provide analogs of a capillary barrier on the order of 10,000 years. Analogs can also assist in characterizing surrounding systems by imitating ecology and high evapotranspiration of diverse native vegetation. However, analogs must be used with caution, since unknown analogs may not have survived into the present, and prior environmental exposure may not be known.

- Recommendation to design for catastrophic events. For example, if investigations have shown catastrophic scouring of the valley in the past, then the disposal site should be sited out of the valley instead of in the valley.

- Improve communication between developers of numerical models and those who provide model support. These communications should be documented in the quality assurance/quality control (QA/QC) program. The QA/QC pedigree must be robust.

- Soil development will be a function of climate, parent material, topography, biology, and time. Pedogenic features will affect hydraulic processes and plant communities at various scales, both spatially and temporally. Analysis of the modern landscape provides a record of soil evolution and pathways that can support long-term assessments of covers.

- Performance monitoring allows for a greater understanding of total system performance, in contrast to compliance monitoring which simply reveals a component of the system as a pass or fail.

- Continue to improve model support and confidence building activities. This process should not cease after acceptance. Re-evaluate the overall facility's performance every 5 to 10 years to ascertain that the design and construction match expected performance.

7.4   Extended Abstracts

The workshop organizing committee considered the topic of model support to be a central issue in evaluating engineered barrier performance. Following the review of technical literature on model support for performance assessments of waste facilities, the workshop organizing committee discussed relevant model support studies. Invited speakers from industry, DOE, NRC licensing staff and academia provided the following extended abstracts. These papers focus on experiences in model support that provide multiple lines of evidence to gain confidence in long-term performance.

7.4.1 Overview of Model Support (for Engineered Barriers)

David W. Esh

Office of Federal and State Materials and Environmental Management Programs
U.S. Nuclear Regulatory Commission
david.esh@nrc.gov

Models are increasingly being used to estimate and project future risks from a variety of facilities associated with the use of nuclear materials. Model support is a critical element in the application of engineered barriers to limit risk from low-level waste disposal facilities, decommissioning sites, and uranium mill tailings facilities. In this abstract, types of support, technical considerations, the use of analogs, and performance confirmation are discussed.

In applications for active systems, model validation is typically required. For active systems, the performance metrics can be observed and used to validate the estimated performance. In practical terms, model validation is the process of ensuring that model predictions provide reliable estimates of future performance. For applications that involve projections into the future and in particular the distant future, validation in the traditional sense is not possible. The use of engineered barriers for waste disposal facilities and decommissioning sites would fall more into this latter class, and therefore to convey reasonable expectations and limit confusion the term model support is used.

Model support is the use of multiple lines of evidence to provide confidence that the projected future performance of the system will reasonably approximate actual future performance. Many engineered barriers are not amenable to model validation in the true sense, therefore model support is expected. Multiple lines of evidence are recommended since in many cases the supporting information will not be direct observation of the output metrics of concern, but rather will likely be indirect observations of intermediate metrics. Use of multiple lines of evidence should reduce the likelihood of errors associated with interpretation of incomplete and inferential information. Model support for performance assessment models was discussed relatively recently in NUREG-1854 (NRC, 2007).

Common principles and good practices that should be considered when developing model support for the use of engineered barriers include:

- Multiple lines of evidence are preferred,
- Direct observations are preferred over indirect observations,
- The level of model support should be commensurate with the risk significance,
- Model support for barriers with a longer experience base can be more limited,
- For very long-term performance, natural analogs should be considered, and
- Support should encompass the full range of expected future boundary and exposure conditions.

Types of Support

Model support can come in many different forms, including but not limited to: analogs, laboratory experiments, field experiments, formal and informal expert judgment, engineering calculations to demonstrated reasonableness of the results (e.g. hand calculations when numerical models are used), alternative model calculations, monitoring data, and comparison to the impacts from past activities. The type of model support will be specific to a particular application. Model support for the long-term performance of engineered systems will be inferential because of the projection in time. The following text provides some relevant examples for engineered covers:

Analogs: Analogs have been used in a variety of programs to help develop support for long-term performance estimates (Waugh, 2004; Simmons, 2002). Analogs for engineered covers were discussed in NRC's guidance for use of engineered barriers in decommissioning sites (NRC, 2003). The greatest uncertainties in the use of engineered barriers stem from extrapolating the results from short-term tests and observations to long-term performance. Analogs may provide valuable insights as to the possible long-term changes to engineered systems.

Laboratory experiments: Laboratory experiments are traditionally used to estimate property values (such as measuring the hydraulic conductivity of a resistive layer such as clay) or provide other local measurements to support engineered barrier performance estimates. Laboratory experiments can allow for isolation and study of individual degradation mechanisms. For example, a column test could be used to estimate the plugging rate of material used for a drainage layer.

Field experiments: Field experiments can provide valuable insights at relevant scales that include expected interactions. Well-designed field experiments are one of the best forms of model support, because they can best represent expected boundary conditions and include interactions. The primary disadvantages are that field experiments can be costly to implement and they are usually only operated for a limited period of time (relative to the desired period of performance). A variety of groups and agencies have sponsored research of engineered covers and some of those activities were summarized in NRC, 2003.

Monitoring data: Monitoring data can provide direct evidence to the performance of a system. However, the monitoring data is usually not available until after the performance of the facility has been assessed and the decision to implement the facility or action has been taken. Monitoring data from nearby analogous facilities can provide indirect evidence. NRC has sponsored research on the development of monitoring systems using performance indicators (NRC, 2007a).

Past Activities: Past activities can provide information as to how systems may respond to future changes in conditions (e.g., recharge, vegetation, land use). Though indirect, the information can be used to support models of engineered cover performance.

Expert judgment: Expert judgment by qualified independent analysts can increase confidence that model projections will be consistent with actual performance. Experts need adequate time to review materials and models used to estimate facility performance. NRC has developed guidance for the use of expert elicitation (NRC, 1996).

Engineering calculations: Engineering calculations can be used to provide confidence in more complex model calculation results by showing how the model estimates are consistent with estimates produced from consideration of well-understood simple physical and chemical processes.

<u>Alternative model calculations:</u> Use of alternative models, without additional lines of evidence, can be problematic because of biases in interpretation of data can influence model parameterization and model selection. Agreement between models does not necessarily mean that either model provides a reliable estimate of future performance. However, use of alternative models combined with other forms of model support, can increase confidence that the model estimates are reasonable.

Technical Considerations

There are a variety of technical considerations to consider when developing a program for model support. Model validation is typically not possible for waste disposal applications, primarily because of the timeframes involved. In addition, the systems and boundary conditions can be complex making direct observation not technically or economically practical. Model support for waste-related applications will typically involve shorter-term and less-direct information, and therefore there is additional uncertainty associated with interpretation. Key technical considerations include but are not limited to: lines of evidence, representativeness, confirmation bias, and inference/causation.

Use of multiple lines of independent evidence can reduce the likelihood that the estimated performance deviates significantly from the actual performance. As more lines of evidence are used, such as those listed above, the probability that each source of information is incorrect decreases. If the information sources conflict, it would typically lead to additional analysis and evaluation. With a single type of support, it is much more likely that a Type II error will be made (the hypothesis (e.g. the model support agrees with the model) is inappropriately accepted).

Ensuring that model support is representative can be a challenge in waste management problems for a variety of reasons. Boundary conditions can be complex and dynamic, varying spatially and temporally. Processes determining overall barrier performance can occur on micro-scales whereas ultimate overall behavior is needed on a macro-scale. Upscaling, or the translation of micro-scale properties and observations to the macro-scale, is especially difficult when the information is sparse and highly uncertain and variable, as it is in many waste management systems. Model support must cover the expected range of conditions that may be imposed on the engineered barriers. The model support information must be representative of those expected conditions.

Confirmation bias is the tendency to prefer information that supports hypotheses. In the context of engineered barriers, this can result in the misinterpretation of sparse and highly variable model support information. Independent evaluation of model support information and its interpretation can help reduce the impact of confirmation bias. In addition, greater amounts and a diversity of model support information can reduce the likelihood of misinterpretation.

One of the biggest challenges with developing model support for the long-term performance of engineered barriers is the lack of direct information for the long timeframes over which performance data may be needed. In many instances, the information that is available may be sparse and highly variable, and the output metric of concern (e.g. the release rate of a radionuclide) may be unobservable and need to be inferred from observation of intermediate outputs. For example, observation of liquid saturation changes may be used to infer liquid flow rates and justify estimated release rates. Release rates can be influenced by the chemistry of the system as well as the physical transport mechanisms. The more removed the model support information is from direct observation of the performance metric, the greater the chance that inferences will be incorrect. Designing a model support program to develop the most direct information as practical is a preferred practice and increase confidence in long-term performance.

Analogs

For some types of engineered barriers, natural analogs may provide information on possible long-term changes to an engineered barrier, such as an engineered cover. Analogs are essentially long-term, uncontrolled field experiments. Evidence from natural analogs can help demonstrate that there are real world complements to the postulated numerical predictions. Analog information is uncertain for a variety of reasons, such as unknown past environmental conditions. Therefore, analog information should not be envisioned as providing proof of future engineered system performance, but rather it provides confidence that the engineered system is likely to perform as designed. The Department of Energy provided a summary of a variety of analogs used to help support the estimated performance of a variety of components of their modeling for the proposed high-level waste repository at Yucca Mountain, Nevada (Simmons, 2002).

Performance Confirmation

In addition to analogs, development of a performance confirmation program can be used to reduce uncertainty and confirm actual performance. Performance confirmation can be used to confirm the data, assumptions, and analyses used to estimate performance of engineered systems such as engineered covers. Performance confirmation would typically involve the monitoring of surface and subsurface conditions associated with the performance of engineered barriers. In addition to monitoring, a performance confirmation program may also include future laboratory and field experiments. One of the biggest challenges associated with a performance confirmation program is to ensure the programs long-term support and maintenance. When an engineered system is first implemented (assuming there were no significant issues with the design or its implementation), the performance should be as expected. This may lead to decision makers abandoning the program. However, the performance (if a deviation from estimated performance is to occur) is more likely to deviate as time progresses.

References

NRC, "Consolidated NMSS Decommissioning Guidance: Characterization, Survey, and Determination of Radiological Criteria," NUREG-1757, Vol. 2, September 2003, ADAMS Accession No. ML032530405.

NRC, "NRC Staff Guidance for Activities Related to U.S. Department of Energy Waste Determinations," NUREG-1854 (Draft Final Report for Interim Use), August 2007, ADAMS Accession No. ML072360184.

NRC, "Integrated Ground-Water Monitoring Strategy for NRC-Licensed Facilities and Sites: Logic, Strategic Approach and Discussion," NUREG/CR-6948, Volume 1, November 2007a.

NRC, "Branch Technical Position on the Use of Expert Elicitation in the High-Level Radioactive Waste Program." NUREG-1563, November 1996.

Simmons, A. et al., "Natural Analogs Synthesis Report," TDR-NBS-GS-000027 REV00 ICN02, Bechtel SAIC Company LLC, Las Vegas, NV, May 2002.

Waugh, W.J., "Design, Performance, and Sustainability of Engineered Covers for Uranium Mill Tailings,' Proceedings of Long-term Performance Monitoring of Metals and Radionuclides in the Subsurface: Strategies, Tools, and Case Studies. U.S. Environmental Protection Agency, U.S. Department of Energy, U.S. Geological Survey, Nuclear Regulatory Commission, April 21-22, 2004, Reston, VA, 2004.

7.4.2 Activities that Support the Scientific Credibility of Radioactive Waste System Performance Models

Abraham Van Luik

U.S. Department of Energy
Office of Environmental Management
Carlsbad Field Office
Carlsbad, NM

Regulations applied to the safety evaluations (performance assessments) for the Yucca Mountain repository after final closure, for 10,000 years, included this requirement at 10 CFR 63.114 "A performance assessment used to demonstrate compliance must: . . . (7) Provide the technical basis for models used to represent the 10,000 years after disposal in the performance assessment, such as comparisons made with outputs of detailed process-level models and/or empirical observations (e.g., laboratory testing, field investigations, and natural analogs)."

This requirement can be interpreted to mean that models must be supported by technical information and scientific observation, which is another way of requiring model validation. Model validation in the classic sense of predicting and verifying against an outcome is not possible for any long-term forward projection. What is possible is to build confidence through model testing and comparisons of aspects of the models with known, analogous or otherwise relevant scientific information (see Eisenberg et al. 1999).

For a system that isolates lower-radioactivity-level wastes, the volume of material needed to support calculations of performance would be much less than is expected for a high-level waste facility, but the supporting activities would be broadly similar. When the system is not a first-of-a-kind, is not unique, there is opportunity to share approaches, and even models and scientific study results for similar structures and facilities. There are past, present, and future, as well as domestic and international, radioactive waste facilities whose safety assessments can serve as analogues for each other. There is no need to re-invent what already exists, but in a "nuclear-safety culture" there is a need to continually question and seek to improve what exists.

Broadly applicable model-support activities used for the Yucca Mountain safety evaluations (for 10,000 years and a million years) are described in detail in SNL (2008). These model-support activities were divided into two categories with considerable feedback and overlap: (1) during-development confidence-building activities and (2) post-development confidence building activities. As models are developed they are subjected to verification, stability testing, and uncertainty characterization reviews. What is done after model development is complete and "production" safety evaluations are being done can include several of the following: (2 a) corroboration with auxiliary analyses, (2 b) corroboration with natural analogue information, (2 c) technical review, and (2 d) performance confirmation monitoring and testing.

Auxiliary analyses, in the case for Yucca Mountain safety, included performing detailed deterministic single realization analyses to gain insight and seek errors, and also comparing final results with results from an independently designed and run performance assessment, and performing a different analysis with key uncertainties removed to gain insight into the conservatisms built into the model being used to demonstrate regulatory compliance. In cases where the system being evaluated is analogous to an existing operating system or a previously utilized but now closed system, domestic or international, comparison against safety evaluations for those analogous systems could help build confidence.

Analogues, of both the natural and anthropogenic variety, can be very useful. They can allow comparisons of materials and conditions relevant to the time-frames of interest in radioactive waste disposal systems. Their study and comparison can provide insights and build confidence in the conceptual and numerical models used to represent processes and events relevant to the system being evaluated.

Several analogues were important in providing support for the Yucca Mountain safety evaluation, but these examples may not be very useful for systems in the near-surface. A potentially useful analogue may be the infiltration of precipitation into different natural (or "engineered") structures and materials. Rock-shelters and caves have been studied in terms of the annual rainfall and amount of infiltration at depth, and determinations of effective percolation have been estimated for ancient burial sites.

Independent technical review can (1) lead to corrections and improvements in a complex model, and (2) can help outside persons and organizations, including societal decision-makers, assess whether or not they have confidence in the safety evaluation and its output. An international peer review was conducted during the Site Recommendation phase of the Yucca Mountain project. One of its conclusions was cited by the Secretary of Energy in his letter to the President recommending site-approval.

Performance confirmation is a program of monitoring, tests, experiments, and analyses conducted to evaluate the adequacy of the information used to demonstrate safety
A performance confirmation program should demonstrate that the system and the sub-system components (i.e., barriers) are operating as anticipated via monitoring

References

SNL (Sandia National Laboratories), 2008, Total System Performance Assessment Model/Analysis for the License Application, MDL-WIS-PA-000005 Rev 00, AD 01, U.S. Department of Energy Office of Civilian Radioactive Waste Management

Eisenberg, N.A.; Lee, M.P.; Federline, M.V.; Wingefors, S.; Andersson, J.; Norrby, S.; Sagar, B.; and Wittmeyer, G.W. 1999. Regulatory Perspectives on Model Validation in High-Level Radioactive Waste Management Programs: A Joint NRC/SKI White Paper. NUREG-1636. Washington, D.C.: U.S. Nuclear Regulatory Commission

7.4.3 Geomembrane Performance in Landfill Cover Systems

George R. Koerner Ph.D., P.E. & CQA, gkoerner@dca.net

Geosynthetic Institute, 475 Kedron Ave. Folsom PA USA
(610) 522-8440
www.geosynthetic-institute.org

Durability is an issue for all construction materials, including geosynthetics, when long design lifetimes are required. Geosynthetics have been shown to exhibit service life of over 100 years in laboratory studies (Lord 1984, Haxo 1985, Hsuan 1998, Koerner1990 and Rowe 2002). Geosynthetics are formulated materials consisting of, (i) the resin from which the name derives, (ii) carbon black or colorants, (iii) short-term processing stabilizers, and (iv) long-term antioxidants. Commonly used geosynthetic polymers include but are not limited to PE, PP, PVC, PET, PA and PS. However for the purpose of this discussion we will concentrate on polyolefins (in particular HDPE and LLDPE) because they are the commodity polymers most likely to be used in this long term landfill cover applications.

When discussing long term service life, various degradation mechanisms can act in isolation or synergistically. Mechanisms such as ultraviolet radiation, temperature and or oxidation have been discussed at length in the open literature, (Schnabel 1981, Whyatt 1990 and Hsuan 1998.) It should be mention that radiation will degrade polyolefins if exposure is at a sufficiently high intensity and over a long enough period of time. The ionic bond strength of most commercially available polyolefins is not high enough to inhibit the degradation process and prevent chain scission beyond low level waste. There are a number of references on the effects of radiation on polymer properties (Charlesby 1960, Carlson, Chnela and Lacoste 1991 and Nimitz et. al. 2001. The effects of γ-rays, neutrons and β-rays are essentially equivalent when their different penetrating powers are considered. The properties of a polyethylene start to change at a total radiation dose of between 10^6 and 10^7 rads (Edy 1979). For reference purposes, the lethal dose of radiation to a human is 100 to 1,000 rads. Whyatt and Fansworth (1990) have evaluated a number of different geomembranes in simulated short-term tests in a high pH (≈ 14 weight percent NaOH) inorganic solution at 90°C and subjected them to radiation up to 39×10^6 rads. It was found that only polyolefin geomembranes were unaffected by the radiation at this level. Furthermore, the radiation did not have a significant effect on other chemical degradation rates. Hence, only polyethylene geomembrane will be called upon as a barrier material in a cover system for the purpose of this discussion. It is interesting to note that Mansour (1974) showed at length that LDPE outperforms HDPE when exposure to high-energy irradiation. He showed that radiation causes cross linking in the polymer which results in a density increase and thus loss of ductility. Since LDPE starts as a more amorphous, less crystalline, polymer than HDPE it is less likely to lose its ductility after prolonged exposure.

In the absence of double lined cover system data, we reference base liner system data to illustrate field geomembrane performance. Note that the hydraulic heads are greater on liners than on covers therefore, better performance is expected for a geomembrane used in a cover system. Two sets of performance data are presented in Figures 1 and 2. The first from the U.S.

EPA, (Othman, Bonaparte and Gross, 1997) which show the performance of 199 landfills over several life cycle stages. It is interesting to note that the leakage rate of the GM/GCL liner system after final cover placement is near zero. Considering the average primary leakage rates from 34 NYS Landfills shown in Figure 2, we can see that the average secondary flows are 7 gpad. This means that the upper liner systems are operating at efficiency over 99%. These two separate studies corroborate the good performance of composite liner systems. They serve as direct observations of containment efficiency when using geosynthetic liner systems.

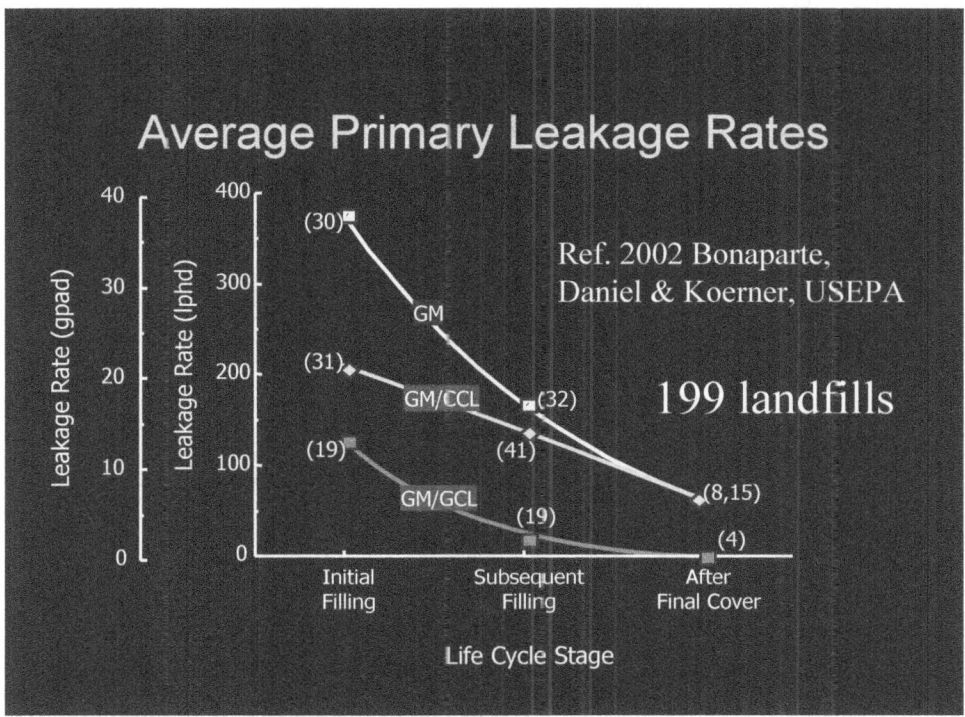

Figure 1- Average Primary Leakage Rates from 199 USA Landfills

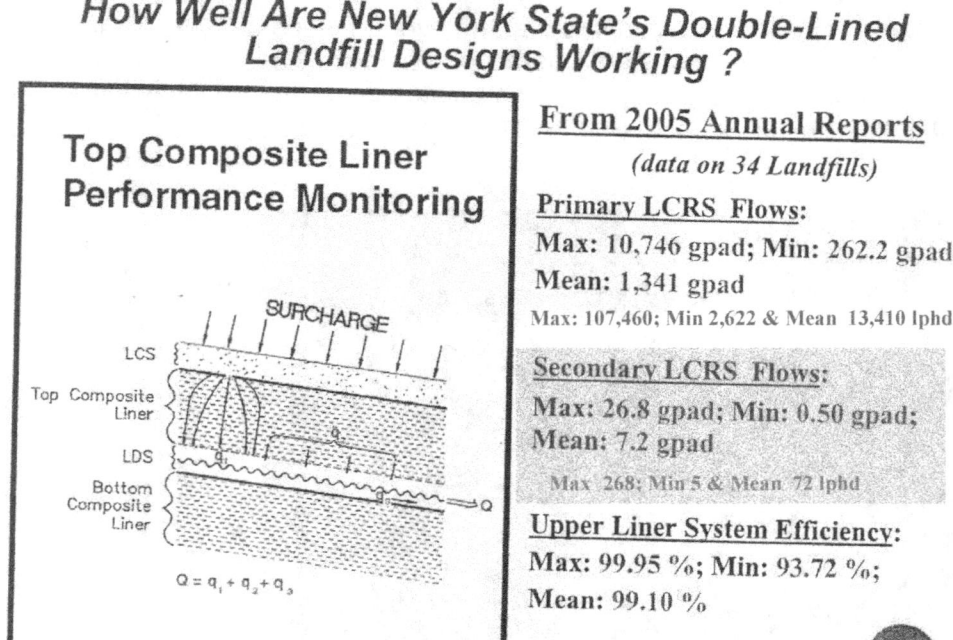

Figure 2- Average Primary Leakage Rates from 34 NYS Landfills

There are a number of papers presenting case histories of geomembrane cover performance, (Youngblood 2009, Case 2010 and Yako 2010). In most cases, original property values are contrasted to the aged material values. Test results indicate a slight decrease in density and oxidative induction time. However, all other physical and mechanical properties have remained essentially unchanged over time periods of up to 30 years, (note that 1980 was the beginning of PE geomembrane manufacturing in the USA.) There are no visible indications of geomembrane degradation and all have performed well over time.

The majority of the long term durability research has been on buried HDPE used in landfill base liner applications. While this material promises service lifetime of hundreds of years when covered, the elevated temperature and UV radiation of exposure will greatly compromise its service life. However, it should be noted that many exposed geomembranes are still in use after 25 years. To be noted, however, is that degradation is a very slow process and extended service life is predicated on good raw materials, an effective formulation, excellent quality control, good design, quality installation and diligent post closure maintenance. In summary, it is strongly believed that if one desires an environmentally safe and secure final cover for a landfill it should include a polyethylene geomembrane or a geosynthetic clay liner, and preferably both in the form of a composite barrier.

References

Carlson, D.J., S. Chnela and L. Lacoste, In Prediction of Effects on Polymers: Editors Clough R. L. and Shalaly S. W. ACS Symposium Series 475, ACSB Department., Washington, D. C> 1991 pp. 432.

Case, M. S., Case history of a 16-year old exposed HDPE geomembrane cover, IGS Conference Brazil, 2010 pp. 27.

Charlesby, A., "Polyethylene," *Atomic Radiation and Polymers*, Pergamum Press, NY, 1960, pp. 198-257.

Eby, R. K. "Durability of Macromolecular Materials," American Chemical Society Symposium Series, No. 95, American Chemical Society, Washington, DC, 1979, 476 pp.

Hsuan, Y. G. and Koerner, R. M., "Antioxidant Depletion Lifetime in High Density Polyethylene Geomembranes," Jour. Geotech. And Geoenviron. Engr., ASCE, Vol. 124, No. 6, 1998, pp. 532-541.

Haxo, H. E., Nelson, N. A. and Miedama, T. A., "Solubility Parameters for Predicting Membrane Waste Liquid Compatibility," Proc. EPA Conf. on Hazardous Waste, Cincinnati. OH, Apr. 1985, pp. 198-215

Koerner, R. M., Designing with Geosynthetics, 5th Ed., Prentice Hall Publ. Co., Englewood Cliffs, NJ, 2005, pp. 2005.

Lord, A. E. and Koerner, R. M., "Fundamental Aspects of Chemical Degradation of Membranes," Proceeding International Conference on Geomembranes, Denver Colorado, June 1984. pp. 293-298, Published by IFAI, St. Paul, Minnesota.

Mansour, S. A., "Effect of gamma irradiation on the Mechanical Relaxation Behavior of High and Low Density Polyethylene," Egypt Journal Sol., Vol. (24) No. (1) 2001, pp. 89-100

Nimitz, J. S., R. E. Allied and B. W. Gordon, "Chemical Compatibility Testing" SANDIA Report, SAND 2001-1998 pp. 91

Othman, M. A., Bonaparte, R. and Gross, B. A. (1997), "Results of Study of Composite Liner Field Performance," Proc. GRI-10 Conference on "Field Performance of Geosynthetics and Geosynthetic Related Systems," R. M. Koerner, G. R. Koerner and Y. G. Hsuan, Eds., GII Publications, Philadelphia, PA, pp. 115-142. (Final Report to be published by U. S. EPA, Cincinnati, OH, 2002)

Rowe, R. K. and Sangam, H. P. (2002), "Durability of HDPE Geomembranes," Jour. of Geotextiles and Geomembranes, Vol. 20, No. 2, April, pp. 77-96.

Schnabel, W., Polymer Degradation: Principles and Practical Applications. Macmillan Publishing, New York 1981 pp.532.

Whyatt, G. A. and Fansworth, R. K., "The High pH Chemical and Radiation Compatability of Various Liner Materials," Geosynthetic Testing for Waste Containment Applications, ASTM STP 1081, Robert M. Koerner, editor, ASTM Publications, Philadelphia, PA USA 1990

Yako, M. A, Case history of a 20-year old exposed HDPE surface impoundment liner, IGS Conference Brasil, 2010 pags 4.

Youngblood, J. H. "Long-Term Field Performance of HDPE Geomembranes in Landfill Applications " ASTM workshop, GRI-23 Conference on "Field Performance and Durability of Geosynthetics and Geosynthetic Related Systems," San Antonio, Texas, 2009 pp. 42.

7.4.4   A Role for Natural Analogs in the Design and Long-Term Performance Evaluation of Earthen Covers for Uranium Mill Tailings

William J. Waugh

S.M. Stoller Corporation
jody.waugh@lm.doe.gov

Introduction

Earthen covers for uranium mill tailings disposal cells are designed to limit potential release pathways including radon gas emanation, water percolation, erosion, and, in some cases, biointrusion. Conventional covers rely on a layer of densely compacted, fine-grained soil as a barrier for radon flux and water percolation, and on rock riprap for erosion control (DOE 1989). Alternative covers rely on (1) a thick, fine-grained soil layer to protect the underlying compacted soil layer from frost and desiccation and to serve as a water storage layer, (2) evaporation and plant transpiration to seasonally remove stored water, and (3) a surface layer of rocky soil for erosion protection (Richardson and Waugh 1996; Waugh et al. 2009).

Earthen covers are engineered ecosystems. The designs are essentially manipulations and amendments of local soils and biology. Changes in soils and biology place covers in a state that is dissimilar to and greatly influenced by surrounding ecosystems. Ecological succession and soil development processes in covers, linked to the surrounding environment and modulated by climatic variability, may fairly rapidly alter engineered soil properties, biological structure, and ecological function in ways that can compromise cover performance (Waugh et al. 1999; Waugh et al. 2007; Benson et al. 2010). Natural analogs may provide clues to help designers build more sustainable covers that mimic favorable natural systems, to help analysts identify scenarios for modeling long-term changes and performance of covers, and to help site managers identify and monitor leading indicators of change. Natural analogs may also help demonstrate to the public that numerical predictions can be supported by existing physical evidence.

This paper (1) gives an overview of the influences of climate change, soil development, and ecological succession on the performance of both conventional and alternative covers, (2) presents examples of natural analogs for improving cover designs and for evaluating long-term performance, and (3) reviews a general performance evaluation process that combines natural analogs with monitoring and modeling.

Analogs of More Sustainable Covers

Evaluations of natural analogs (sometimes called reference areas) can provide a target for designing more sustainable covers. An evaluation of relatively undisturbed soils similar to the borrow source for a cover can provide a target for revegetation and evidence of how soil development processes may influence soil physical and hydrological properties (Albright et al. 2010). Evaluations of analog sites can also be used to develop design and performance criteria for cover soils and revegetation. If the goal is to mimic the favorable ecology and soil water balance of an analog site, achieving the target soils and ecology may take many years. As-built soil hydraulic properties, including saturated hydraulic conductivity in covers, should be as close as possible to the properties of reference area soils. Hydraulic conductivity of densely compacted soils can increase and become consistent with reference areas within only 5–10 years (Benson et al. 2010). However, soil edaphic properties (soil properties related to plant growth) and soil morphology—the complex exchange of energy and nutrients among plants, animals, microbes, soils, and topography—may take tens to hundreds of years to become like the analog soil. Establishment of the reference target for plant species composition might be achieved in a few years if propagules from planted seeds and transplants survive and reproduce; however, these young plants may take years to decades (for grasslands and shrublands) and centuries (for woodlands and forests) to mature before revegetation targets for plant community structure (e.g., height, cover, leaf area, rooting depth) can be achieved (Albright et al. 2010).

The Hanford site in Washington is an example of a natural analog that can improve design concepts. An undisturbed soil profile near the site consists of a thick, fine-textured soil layer overlying a coarse gravel layer that functions as a natural capillary barrier. The soil profile formed in sediments that were rapidly laid down approximately 13,000 years ago near the end of a period of Pleistocene cataclysmic floods (Baker et al. 1991). The profile is unique in that the entire sediment sequence was laid down, and pedogenic (soil development) processes started forming at a known time and then continued, relatively uninterrupted, until the present.

As an analog, the Hanford soil profile provides clues about the integrity of layer interfaces and the potential performance of a capillary barrier cover at the site (Bjornstad and Teel 1993). An open-work gravel layer suggests that a capillary barrier could be designed that prevents illuviation of fine soil into underlying clean gravel, preserving the integrity of the layer interface for thousands of years. Thick calcium carbonate deposits at the fine layer–coarse layer interface is a pedogenic clue that the capillary barrier has limited deep percolation at this site for many centuries; however, calcium carbonate deposits on gravels below the interface suggest that water sometimes percolates past the capillary barrier and, hence, the fine soil layer may not be thick enough to provide adequate water storage for all precipitation events.

Long-term Performance Analogs

Evaluations of natural analogs can provide clues to better understand inevitable changes in the cover environment and to develop possible scenarios for modeling long-term performance of covers (Waugh et al. 1994). Evidence from natural analogs can improve our understanding of (1) meteorological variability associated with possible long-term changes in climate; (2) vegetation responses to climate change and disturbances; (3) effects of plant community dynamics on evapotranspiration, soil permeability, soil erosion, and animal burrowing; and (4) effects of soil development processes on water storage, permeability, and site ecology. Examples follow of natural and archaeological analogs for waste disposal sites that were characterized to discern possible long-term changes in environmental settings. Examples are given for climate change, pedogenesis (soil development), and ecological succession.

Climate data are required for designing and projecting long-term performance of engineered covers (Ho et al. 2004). Projections of long-term extreme events and shifts in climate states over hundreds and thousands of years may be required, as well as annual and decadal variability in meteorological parameters. Methods have been demonstrated that combine global change models and paleoecological evidence to establish a first approximation of possible future climatic states at sagebrush steppe sites at the Hanford site in Washington (Peterson 1996) and at the Monticello uranium mill tailings disposal site in Utah (Waugh and Petersen

1996). Paleoclimate data for Monticello indicated an average annual temperature range of 2 to 10 °C and a precipitation range of 80 to 60 cm, corresponding to late glacial and mid-Holocene periods. Instrumental records for regional stations were then used as a basis for selecting soil and vegetation analog sites that span a reasonable range of future climate states.

Pedogenic processes will change soil physical and hydraulic properties that are fundamental to the performance of engineered covers. Although rates and magnitudes of change vary, pedogenesis takes place to some degree in all soils. Pedogenesis includes processes such as (1) formation of macropores for preferential flow associated with root growth, animal holes, and soil structural development; (2) secondary mineralization, deposition, and illuviation of fines, colloids, soluble salts, and oxides that can alter water storage and movement; (3) soil mixing caused by freeze-thaw activity, animal burrows, and the shrink-swell action of expansive clays; and (4) formation of lag layers by winnowing, frost heaving, movement of soil gases during and after rain, and the shrink-swell action of expansive clays (e.g., Chadwick and Graham 2000; McFadden et al. 1998; Collis-George 1991; McDonald et al. 1996).

Soil development analogs were investigated for the Monticello disposal cell. Key soil physical and hydraulic properties were measured in natural and archaeological soil profiles at climate analog sites to infer possible future pedogenic effects on cover performance.

Ecological change is inevitable and may alter the functional performance of all designed covers, those with both conventional and alternative designs, often in ways not anticipated. Plant communities develop and change in response to several interacting factors: propagule accessibility, climatic variability, change in soil characteristics, disturbances such as fire, and species interactions such as herbivory, competition, or fluctuations in soil microbe populations. Plant community dynamics are manifested by shifts in species composition, vegetation abundance, and species diversity and may be accompanied by changes in rates of nutrient cycling, energy exchange, and transpiration. Changes may be salient, such as a conversion from shrublands to grasslands after fire, or more subtle, such as a gradual replacement of seeded perennial grasses by annual grasses, causing a drop in water removal by transpiration, because of competition and inadequate soil fertility.

Even in the absence of large-scale disturbances, seasonal and annual variability in precipitation and temperature will cause changes in species abundance, diversity, biomass production, and soil water extraction rates (Anderson et al. 1993; Link et al. 1994b). It will be important to know how changes in the plant community inhabiting a cover may influence soil water movement, evapotranspiration, and the water balance of a cover. In the long term, the ecology of a cover will change in ways that cannot be accurately predicted by models or short-term field tests. For example, successional changes in the vegetation can create small-scale topographic patterns that foster greater heterogeneity in the soil water balance. At arid sites, desert shrub communities that are likely to develop on covers tend to trap windborne sediments, causing a hummock-swale relief with variable soil physical and hydraulic properties and enhancement of mosaics in the plant community (Link et al. 1994a). Similarly, at humid sites, blowdown of mature trees growing on engineered covers will create depressions for water accumulation (Suter et al. 1993).

Successional chronosequences provide clues of possible future ecological change. For example, at the Lakeview, Oregon, site, possible future responses of plant community composition and leaf area index to fire were evaluated using a nearby fire chronosequence (Waugh 2004). Possible vegetation responses to climate change scenarios were evaluated at regional climate-change analog sites. Leaf area index, as an index of plant transpiration, ranged from 0.15 to 1.28 for the fire chronosequence and from 0.43 to 1.62 for dry and wet climate analog sites.

Long-Term Performance Evaluation Process

One approach for evaluating the long-term performance of disposal cell covers links probabilistic modeling with evidence of long-term change derived, in part, from natural analogs. Natural analogs help define possible future environmental scenarios. This approach can be applied initially during the design phase, with the objective of building more sustainable covers, and then during the maintenance and monitoring phase, to reiterate long-term performance projections. A probabilistic modeling platform developed by Pacific Northwest National Laboratories, called Framework for Risk Analysis in Multimedia Environmental Systems (FRAMES), is an example that has been used for hazardous waste landfills (http://mepas.pnl.gov/FRAMESV1/). Ho et al. (2004) demonstrated applications of FRAMES for evaluations of uranium mill tailings disposal cells near Monticello and Lakeview.

The following general steps describe a systematic approach for projecting long-term performance that links natural analogs, modeling, and monitoring (after Ho et al. 2004).
1. *Develop and screen future environmental scenarios.* A scenario is a well-defined sequence of processes or events that describes possible future conditions of the cover. For example, a scenario might include a future climate state based on global change models, future ecological conditions and stages of soil development for the climate state, and a different land use.
2. *Develop models of relevant future scenarios.* Broad conceptual models of future scenarios are developed first to guide the selection of mechanistic or probabilistic models. Specific models can then be selected and integrated into a total system model framework that links performance with risk, such as FRAMES.
3. *Develop values and uncertainty distributions for input parameters.* Single deterministic values might be assigned to some well-characterized parameters, but uncertainty distributions are preferable. The uncertainty and/or variability in other parameters may require the use of uncertainty distributions to define values. Uncertainty distributions for many environmental values will be based on the characterization of natural analogs. Some uncertainty distributions may be derived from the literature, prototype tests in lysimeters, or monitoring results from landfills in similar environments.
4. *Perform calculations and sensitivity/uncertainty analyses.* If performance calculations (runs) include uncertain parameters, a Monte Carlo approach can be used to rapidly create large suites of simulations that input different combinations of parameter values sampled from the uncertainty distributions. The results are a collection of uncertainty distributions that can be compared to the performance objectives. Sensitivity analyses indicate which input parameters the performance metrics are most sensitive to.
5. *Document results and iterate previous steps as needed.* The results are presented as the probability of exceeding a performance objective. Results can be used to iteratively evaluate alternative designs and components and to select the most suitable cover design for the conditions at a site.
6. *Monitor key performance indicators.* Use results of sensitivity analyses to help select parameters for post-closure performance monitoring of the cover. The objectives of performance monitoring include (1) provide early warnings of possible deterioration of the cover, (2) compare actual performance results with model predictions, and (3) reiterate and refine long-term performance projections, particularly in response to changes in the environmental setting.

References

Albright, W.H., C.H. Benson, and W.J. Waugh, 2010. *Water Balance Covers for Waste Containment: Principles and Practices*, American Society of Civil Engineers, Reston, VA.

Anderson, J.E., R.S. Nowak, T.D. Ratzlaff, and O.D. Markham, 1993. "Managing soil moisture on waste burial sites in arid regions," *Journal of Environmental Quality*, 22:62–69.

Baker, V.R., B.N. Bjornstad, A.J. Busacca, K.R. Fecht, E.P. Kiver, U.L. Moody, J.G. Rigby, D.F. Stradling, and A.M. Tallman, 1991. "Quaternary Geology of the Columbia Plateau," In K.B. Morrison (ed), *Quaternary Nonglacial Geology; Conterminous U.S.: The Geology of North America,* Vol. K-2, Geological Society of America, Boulder, CO.

Benson, C.H., W.H. Albright, D.O. Fratta, J.M. Tinjum, E. Kucukkirca, S.H. Lee, J. Scalia, P.D. Schlicht, and X. Wang, 2010 (DRAFT). *Engineered Covers for Waste Containment: Changes in Engineering Properties and Implications for Long-Term Performance Assessment*, NUREG/CR-XXXX, U.S. Nuclear Regulatory Commission, Office of Research, Washington, DC.

Bjornstad, B., and S. Teel., 1993. *Natural Analog Study of Engineered Protective Barriers at the Hanford Site*, PNL-8840, Pacific Northwest National Laboratory, Richland, WA.

Chadwick, O.A., and R.C. Graham, 2000. "Pedogenic processes," in M.E. Summer (ed.), *Handbook of Soil Science*, CRC Press.

Collis-George, N., 1991. "Drainage and soil structure: a review," *Australian J. Soil Res.*, 29:923–933.

DOE, 1989. *Technical Approach Document, Revision II,* UMTRA-DOE/AL 050425.0002, United States Department of Energy, Albuquerque, NM.

Ho, C., B. Arnold, J. Cochran, R. Taira, and M. Pelton, 2004. "A Probabilistic Model and Software Tool for Evaluating the Long-term Performance of Landfill Covers," *Envir. Modeling and Software*, 19: 63–88.

Link, S.O., W.J. Waugh, and J.L. Downs, 1994a. "Effects of coppice dune topography and vegetation on soil water dynamics in a cold-desert ecosystem," *Journal of Arid Environments*, 27: 265–278.

Link, S.O., W.J. Waugh, and J.L. Downs, 1994b. "The role of plants on isolation barrier systems," in G.W. Gee and N.R. Wing (eds.), *In-Situ Remediation: Scientific Basis for Current and Future Technologies,* Battelle Press, Columbus, Ohio.

McDonald, E.V., F.B. Pierson, G.N. Flerchinger, and L.D. McFadden, 1996. "Application of a Process-based Soil-water Balance Model to Evaluate the Influence of Late Quaternary Climate Change on Soil-water Movement in Calcic Soils," *Geoderma*, 74: 167–192.

McFadden, L.D., E.V. McDonald, S.G. Wells, K. Anderson, J. Quade, and S.L. Forman, 1998. "The Vesicular Layer of Desert Soils: Genesis and Relationship to Climate Change and Desert Pavements Based on Numerical Modeling, Carbonate Translocation Behavior, and Stable Isotope and Optical Dating Studies," *Geomorphology*, 101–145.

Peterson, K., 1996. "Projecting the Range of Potential Future Climate Change as an Aid in Assessing the Effectiveness of the Hanford Site Permanent Isolation Barrier," in S. Ghan, W. Pennell, K. Peterson, E. Rykiel, M. Scott, and L. Vail, eds., *Regional Impacts of Global Climate Change: Assessing Change and Response at Scales that Matter,* Battelle Press, Columbus, OH.

Richardson, G.N. and W.J. Waugh, 1996. "The Design of Waste Repository Final Cover Systems for Arid and Semiarid Regions," pp. 645–658, in H.Y. Fang and H.I. Inyang (eds.), *3rd International Symposium on Environmental Geotechnology*, Technomic Publishing Company, Lancaster, PA.

Suter, G.W. II, R.J. Luxmoore, and E.D. Smith, 1993. "Compacted soil barriers at abandoned landfill sites are likely to fail in the long term," *Journal of Environmental Quality*, 22: 217–226.

Waugh, W., and K. Petersen, 1996. "Climate Change and Uranium Mill Tailings in the Four Corners Region," in S. Ghan, W. Pennell, K. Peterson, E. Rykiel, M. Scott, and L. Vail, eds., *Regional Impacts of Global Climate Change: Assessing Change and Response at Scales that Matter*, Battelle Press, Columbus, OH.

Waugh, W., K. Petersen, S. Link, B. Ejornstad, and G. Gee, 1994. "Natural Analogs of the Long-term Performance of Engineered Covers," n G. Gee and N. Wing, eds, *In-situ Remediation: Scientific Basis for Current and Future Technologies*, Battelle Press, Columbus, Ohio, 379–409.

Waugh, W.J., 2004. "Designing Sustainable Covers for Uranium Mill Tailings," pp 157–173, in W.R. Keammerer and J. Todd, *Proceedings, High Altitude Revegetation Workshop,* No. 16. Information Series No. 99, Colorado Water Resources Research Institute, Colorado State University, Fort Collins, Colorado.

Waugh, W.J., G.M. Smith, B. Danforth, G.W. Gee, V. Kothari, and T. Pauling, 2007. "Performance Evaluation of the Engineered Cover at the Lakeview, Oregon, Uranium Mill Tailings Site," Proceedings of Waste Management 2007 Symposium, Tucson, AZ.

Waugh, W.J., K.L. Petersen, S.O. Link, B.N. Bjornstad, and G.W. Gee, 1994. "Natural Analogs of the Long-term Performance of Engineered Covers," pp 379–409, in G.W. Gee and N.R. Wing (eds.), *In-Situ Remediation: Scientific Basis for Current and Future Technologies*, Battelle Press, Columbus, OH.

Waugh, W.J., S.J. Morrison, G.M. Smith, M. Kautsky, T.R. Bartlett, C.E Carpenter, and C.A. Jones, 1999. *Plant Encroachment on the Burrell, Pennsylvania, Disposal Cell: Evaluation of Long-term Performance and Risk*, GJO-99-96-TAR, U.S. Department of Energy, Grand Junctior CO.

Waugh, W.J., C.H. Benson, and W.H. Albright, 2009. "Sustainable covers for uranium mill tailings, USA: Alternative design, performance, and renovation," Proceedings of 12th International Conference on Environmental Remediation and Radioactive Waste Management, Liverpool, UK.

7.4.5 Arid Soil Evolution and Pedologic Development: Process Considerations and Applications to Engineered Barrier Design

Todd G. Caldwell[1], Michael H. Young[2], Eric V. McDonald[1]

[1]Desert Research Institute, Division of Earth and Ecosystem Sciences, 2215 Raggio Parkway, Reno, NV 89512
[2]University of Texas at Austin, Bureau of Economic Geology, University Station, P.O. Box X, Austin, TX 78712

Soil development in arid regions and the resulting soil morphology are functions of climate, biologic activity and, most importantly, time. Soil development begins when depositional or erosional processes cease and the geomorphic surface becomes stable, transforming surficial sediment to soil (i.e., pedogenesis). Pedogenic features affect a multitude of hydrologic processes operating across a range of spatial and temporal scales relevant to engineered surface barriers (ESB). At the pore-scale (μm-mm), soil development results in the formation of structure and secondary porosity. At the field-scale (m-km), soils co-evolve with climate and vegetation to form a mosaic of vegetation patches and open soil (or interspace). ESB are anthropogenic landforms with design criteria based on static, engineered hydraulic properties that will ultimately evolve through both coupled biotic and abiotic processes. The rate of post-construction soil development and its effect on performance criteria is uncertain (Benson et al., 2007) and highly variable from site to site. Natural analogs may provide some direct lines-of-evidence for long-term evaluation periods, which may exceed 10,000 years for low- and high-level radioactive waste repositories intended for arid lands.

Soil development is often investigated through the use of soil chronosequences - a genetically related suite of soils evolving under similar vegetation, climate and topography with only time varying (Harden, 1982). Chronosequence studies on arid alluvial fans generally assume that soil development is a quasi-linear transition from a geomorphically young soil to a mature soil, developing progressively under the influence of major soil-forming factors (climate, topography, biology, parent material, and time) to reach some pedogenic equilibrium. For example, soils on arid alluvial fan complexes transform from coarse-textured, bar and swale deposits into smooth, desert pavements (McDonald et al., 2003). The primary driver for this pedogenic transformation is the influx and deposition of fine-grained eolian dust (Reheis et al., 1995), enriching alluvial deposits in silt, clay, and soluble salts. Low mean annual precipitation and high evaporative demand result in arid soils being net sinks for both dust and solutes. With the onset of dust accumulation, soil development begins just below the surface with the formation of a vesicular horizon (V, formerly Av), a silt-rich soil horizon ubiquitous to deserts. Infiltration capacity in arid soils is reduced over time (Young et al., 2004), resulting from this abiotic process of dust sequestration (Yaalon and Ganor, 1973; McFadden et al., 1987). As the V horizon develops, infiltration mechanisms shift from predominantly bulk matrix flow, due to the coarse texture and open pore-system, to a macropore dominated system with flow constricted to faces of prismatic, highly structured peds (Meadows et al., 2008). The hydraulic conductivity function $[K(\psi)]$ is affected during this progression, thereby limiting infiltration, increasing water holding capacity (McDonald et al., 1996; Young et al., 2004), and reducing plant available moisture (Hamerlynck et al., 2002). The continued development of the V horizon results in increased surface runoff and an overall decrease in surface stability. Ultimately, the landform is incised and eroded into a ballena (Wells et al., 1987). The transition of stable to unstable landforms is generally driven by climate change or tectonic activity in natural environments during the Quaternary.

In practice, soil forming factors create highly dynamic, non-linear systems of evolution that are often more chaotic (Huggett, 1998). Densely compacted, layered materials associated with ESB are in disequilibrium with the environment and will likely evolve at rates much faster than those observed in more natural pedogenesis (Albright et al., 2006; Benson et al., 2007), although the biotic and abiotic drivers will be analogous. Pedogenic processes can be quantified with moderate certainty and applied to the evolution of ESB and the landforms they will ultimately become. As vegetated evapotranspiration covers gain wider acceptance for closure of waste disposal sites in arid and semi-arid regions, it is useful to view them as landforms that are subject to environmental change over time, with biotic processes working at shorter temporal scales (10s to 100s of years), and abiotic processes requiring 1,000 to 10,000 year time scales. Spatially, biotic processes are typically confined to the vegetation (i.e., meter) scale, while abiotic processes operate at the basin scale. The performance assessment of these anthropogenic landforms must account for both processes operating at different spatial and temporal scales.

Biotic processes have passive and active consequences in pedogenesis. Passively, soil development exerts significant influence over the distribution of perennial desert plants while active bioturbation and nutrient cycling alter undercanopy soil. For example, the depth and lateral spread of creosote (*Larrea tridentata*) roots, a common evergreen shrub of the Mojave Desert, is coupled to the pedology and the depth of water penetration (Stevenson et al., 2009). Young, weakly developed soils have larger canopy volumes and deeper and more laterally extensive root systems than creosote situated on older geologic surfaces with well-developed soils. Numerical simulations and field measurements indicate that percolation depths are limited to approximately 1.5 m on young surfaces and 0.75 m on old interspace surfaces throughout a 50-year simulation of Mojave Desert climate, coinciding with measured rooting depths. Hydraulic conductivity measured underneath canopies (~10 cm h^{-1}) tends to be independent of surface age, because soil mixing by faunal burrowing, bioturbation, root turnover, etc. resets the pedogenic clock. The persistence of soil structure depends on the proximity to plant canopies. Interspace soils can vary from >100 cm h^{-1} on young surfaces to <1 cm h^{-1} on soils with well-developed V horizons (Young et al., 2004; Caldwell et al., 2006; Stevenson et al., 2009). When treated as a binary system of canopy or interspace microsites, deeper water flux is generally observed below plant canopies (Dunkerley, 2002), although nearly all moisture is ultimately scavenged by vegetation. Gradients in hydraulic properties have been observed from canopy to interspace microsites at distances up to 1.4 times the plant diameter (Caldwell et al., 2008). Beyond these distances, the random abiotic processes described above influence interspace hydraulic properties.

This pedologic co-evolution of desert soils leads to smaller shrubs, decreased density, and negative impacts to overall plant performance. The predictable spatial structure in $K(\psi)$ from canopy and to interspace microsites points to an age-dependency of $K(\psi)$. In turn, this knowledge could lead to more accurate parameterization of hydraulic parameter fields based solely on vegetation distributions, when the scale of interest warrants (e.g., a vegetated water balance cover). The coupled effect of vegetation density, soil properties, and soil thickness can further aid in ESB design (Young et al., 2006) but ultimately the coupled abiotic and biotic processes shape and create stable landforms (Shafer et al., 2007). Furthermore, differences in soil structure and texture between undercanopy and interspace microsites can be significant

(Caldwell et al., 2008), leading to a complicated spatial heterogeneity of soil properties that alter the partitioning of evapotranspiration: the most important water removal process operating on arid ESB.

The classic desert pavement soil consists of a finer-grained V and B(w, t, k, km) horizon overtop a coarse grained C horizon which is in many ways similar to capillary barrier design of earthen covers. Ultimately, the use of soil analogs can help benchmark numerical simulations of water and solute transport over the last 20,000+ years, and provide significant information for how soil covers will perform throughout their lifespan, in particular, toward the end of the post-closure period and beyond. A catalog of analogs covering a breadth of climates, parent materials and vegetation sequences could aid in both the design and performance evaluation of ESB. In this way, we can take advantage of the unique spatial and vertical distributions of pedogenic properties (e.g., silt and clay translocation, soluble salts and pavement cover) that provide records of past hydrologic process and dictate current hydraulic and landscape processes. Climatic extremes throughout the Quaternary period drive nearly all pedogenic processes in arid systems, resulting in non-linear (in time) soil structural formation and destruction. These processes and their rates can aid in the long-term assessment of natural surface barriers.

References

Albright, W.H., C.H. Benson, G.W. Gee, T. Abichou, S.W. Tyler, and S.A. Rock. 2006. Field performance of three compacted clay landfill covers. Vadose Zone J. 5:1157-1171.

Benson, C.H., A. Sawangsuriya, B. Trzebiatowski, and W.H. Albright. 2007. Postconstruction changes in the hydraulic properties of water balance cover soils. Journal of Geotechnical and Geoenvironmental Engineering 133:349-359.

Caldwell, T.G., E.V. McDonald, and M.H. Young. 2006. Soil disturbance and hydrologic response at the National Training Center, Ft. Irwin, California. J. Arid Environ. 67:456-472.

Caldwell, T.G., M.H. Young, J. Zhu, and E.V. McDonald. 2008. Spatial structure of hydraulic properties from canopy to interspace in the Mojave Desert. Geophys. Res. Lett. 35, L19406:doi:10.1029/2008GL035095.

Dunkerley, D.L. 2002. Infiltration rates and soil moisture in a groved mulga community near Alice Springs, arid central Australia: Evidence for complex internal rainwater redistribution in a runoff-runon landscape. J. Arid Environ. 51:199-219.

Hamerlynck, E.P., J.R. McAuliffe, E.V. McDonald, and S.D. Smith. 2002. Ecological responses of two Mojave Desert shrubs to soil horizon development and soil water dynamics. Ecology 83:768-779.

Harden, J.W. 1982. A quantitative index of soil development from field descriptions: Examples from a chronosequence in central California. Geoderma 28:1-28.

Huggett, R.J. 1998. Soil chronosequences, soil development, and soil evolution: A critical review. Catena 32:155-172.

McDonald, E.V., L.D. McFadden, and S.G. Wells. 2003. Regional response of alluvial fans to the pleistocene-holocene climatic transition, Mojave Desert, California, p. 189-205, In Y. Enzel, et al., eds. Paleoenvironments and paleohydrology of the Mojave and southern Great Basin Deserts. Geological Society of America Special Paper 368., Boulder, CO.

McDonald, E.V., F.B. Pierson, G.N. Flerchinger, and L.D. McFadden. 1996. Application of a soil-water balance model to evaluate the influence of holocene climate change on calcic soils, Mojave Desert, California, USA. Geoderma 74:167-192.

McFadden, L.D., S.G. Wells, and M.J. Jercinovich. 1987. Influences of eolian and pedogenic processes on the origin and evolution of desert pavements. Geology 15:504-508.

Meadows, D.G., M.H. Young, and E.V. McDonald. 2008. Influence of relative surface age on hydraulic properties and infiltration on soils associated with desert pavements. Catena 72:169-178.

Reheis, M.C., J.C. Goodmacher, J.W. Harden, L.D. McFadden, T.K. Rockwell, R.R. Shroba, J.M. Sowers, and E.M. Taylor. 1995. Quaternary soils and dust deposition in southern Nevada and California. Geol. Soc. Am. Bull. 107:1003-1022.

Shafer, D.S., M.H. Young, S.F. Zitzer, T.G. Caldwell, and E.V. McDonald. 2007. Impacts of interrelated biotic and abiotic processes during the past 125 000 years of landscape evolution in the northern Mojave Desert, Nevada, USA. J. Arid Environ. 69:633-657.

Stevenson, B.A., E.V. McDonald, and T.G. Caldwell. 2009. Root patterns for *Larrea tridentata* in relation to soil morphology in Mojave Desert soils of different ages, p. 312-338, *In* R. H. Webb, et al., eds. The Mojave Desert: Ecosystem processes and sustainability. University of Nevada Press, Reno NV.

Wells, S.G., L D. McFadden, and J.C. Dohrenwend. 1987. Influence of late Quaternary climatic changes on geomorphic and pedogenic processes on a desert piedmont, eastern Mojave Desert, California. Quaternary Res. 27:130-146.

Yaalon, D.H., and E. Ganor. 1973. The influence of dust on soils in the Quaternary. Soil Sci. 116:146-155.

Young, M.H., E.V. McDonald, T.G. Caldwell, S.G. Benner, and D.G. Meadows. 2004. Hydraulic properties of a desert soil chronosequence in the Mojave Desert, USA. Vadose Zone J. 3:956-963.

Young, M.H., W. Albright, K.F. Pohlmann, G. Pohll, W.H. Zachritz, S. Zitzer, D.S. Shafer, I. Nester, and L. Oyelowo. 2006. Incorporating parametric uncertainty in the design of alternative landfill covers in arid regions. Vadose Zone J. 5:742-750.

7.4.6   Historical Studies of Moisture Contents and Infiltration Through Compacted Clay Radon Barriers at Rock Covered UMTRA Disposal Cells

Kent Bostick

Pro2Serve, CGWP, Oak Ridge, TN

Recently, there has been interest in the performance and evolution of Uranium Mill Tailings Remedial Action (UMTRA) Project disposal cell covers. Several UMTRA Project disposal facilities consist of uranium mill tailings covered by one- to two-meter thick compacted clay radon barriers and 15 centimeters (cm) of filter sand, overlain by 30 cm of erosion protection riprap. To comply with U.S. Environmental Protection Agency (EPA) groundwater standards applicable to the UMTRA Project (40 CFR 192), concentration limits for hazardous constituents cannot be exceeded at the downgradient limit of the disposal facility (the point of compliance, or POC). Minimizing infiltration that transports contaminants from the tailings to the uppermost aquifer facilitates compliance with the proposed standards. Radon barriers also have the requirement to reduce radon emanation. Recently, the function and effectiveness of the rock covers has been questioned with regards to changes in the hydraulic properties of the radon barrier related to soil genesis, freeze-thaw cycles, and plant succession. This extended abstract reviews two former UMTRA documents contributed by the author that report on field studies on measuring long-term moisture content and modeling that predicted infiltration rates in the covers in response to semi-arid climatic conditions in the western United States.

Prior to promulgation of the EPA groundwater protection standards for the UMTRA Project, the typical UMTRA Project radon barriers in rock-covered disposal cells were constructed of low-permeability compacted clay with a saturated hydraulic conductivity on the order of 10^{-7} centimeters per second (cm/s). Early estimates of the infiltration flux through radon barriers were conservatively based on assumptions of a unit gradient and a saturated radon barrier. Therefore, the downward flux was equivalent to the saturated hydraulic conductivity of the radon barrier. However, if the radon barrier is continuously unsaturated, its operational hydraulic conductivity may be several orders of magnitude lower, yielding a lower steady state seepage flux from the tailings. If this is the case, the disposal cell will more likely meet the proposed EPA groundwater standards at the POC. This hypothesis needed to be tested to allow approval of the UMTRA cover designs by the Nuclear Regulatory Commission.

Shiprock UMTRA Cover Field Study

A field study was conducted to evaluate moisture conditions in the UMTRA disposal cell cover at the Shiprock, New Mexico, site in the late 1980s. Limited field data also were obtained for the UMTRA disposal cell covers at the Clive, Utah, and Burrell, Pennsylvania, sites. The field study was initiated by the U.S. Department of Energy (DOE) to determine whether the rock-covered tailings disposal cell could continue to be used as a design that would comply with the EPA groundwater protection standards.

Percent saturation profiles were developed for the clay radon barriers at the three disposal cells and capillary moisture curves and unsaturated hydraulic conductivity curves were developed for the Shiprock radon barrier. The radon barriers of all three disposal cells were found to be unsaturated, with the average percent saturation ranging from 82 to 84 percent. As part of the field study, the Shiprock disposal cell was instrumented to monitor meteorological stresses, relative soil tension, and moisture content profiles in the filter layer and radon barrier. Geotechnical samples collected from eight boreholes in the Shiprock radon barrier indicated that the average percent saturation of the radon barrier was 83.6 percent (12.5 percent by weight). The construction moisture content of the radon barrier was 14.9 percent by weight, indicating some drying of the radon barrier may have occurred since placement. Results of neutron logging of the radon barriers indicated that moisture contents were uniform with depth and consistent for the period of monitoring.

Seasonal field evaporation experiments using microlysimetry demonstrated that the potential evaporation from the filter layer exceeds the annual precipitation at the Shiprock site, and evaporation by advection of moisture through openings in the rock layers may be the primary mechanism for removing excess water from the filter layer. The measured evaporative flux was greatest in the summer, at an equivalent flux of 43 cm per year, and was lowest during the fall and winter, at an equivalent flux of 23 cm per year. For comparison purposes, the net annual precipitation at Shiprock is only 15 cm per year. Therefore, evaporation through the rock removes much of the excess moisture in the filter layer (resulting from precipitation) and the filter layer remains unsaturated throughout most of the year. Removal of excess water by evaporation is more significant than lateral runoff in the filter layer. Lateral runoff is minimal, except during the spring, when the saturated filter layer thaws. Runoff from the Shiprock cover is reduced by the low slope of the cover (two percent) and the relatively low saturated hydraulic conductivity (8×10^{-4} cm/s) of the filter layer.

Monitoring of relative soil tensions with time indicated that relative soil tensions in the filter layer and upper portion of the radon barrier are controlled by meteorological stresses. Relative soil tensions in the filter layer decreased during winter, but were generally high the remainder of the year, except after precipitation events. Relative soil tensions were highly variable in the upper portion of the radon barrier, but remained constant below a depth of 60 cm.

To investigate long-term analogs of the performance of thick compacted clay barriers, the DOE performed a literature review on four small, earth-filled, flood-control dams in New Mexico ranging in age from 15 to 27 years. While they were not rock-covered at the time they were studied by the U.S. Soil Conservation Service (SCS), moisture characteristics measured during that study are relevant to the UMTRA Project. The dams are on small drainage basins and impound water only during occasional heavy runoff. The SCS provided data comparing percent saturation versus depth that had been collected to address concerns over surface cracking.

Profiles of percent saturation for four earth-fill dams averaged about 80-90 percent. The maximum percent saturations for each of the dams tend to occur at middle depths in the profiles. The relatively low percent saturations found at depth in several of the profiles are from the more permeable foundation strata below the base of the dams. Lower percent saturations in upper parts in some of the profiles are inferred to result from evaporation; surficial desiccation cracks in the dams appear to have facilitated evaporation to depth of 20 ft. The SCS's aim in studying the dams was to determine the cause of cracking. After ascertaining that the dams were drying near the surface, the SCS remedied the problem by adding rock mulch to their upper surfaces. Because of differences between the dams at the time of sampling and typical UMTRA Project cover designs, (specifically the dams had no rock mulch on their top surfaces) the comparison of moisture characteristics is limited to the observation that these semiarid-zone dams are unsaturated.

Modeling of Infiltration Though the Shiprock UMTRA Cover

Long-term moisture contents in the Shiprock radon barrier were simulated using the finite element unsaturated flow model UNSAT2. The modeling demonstrated that soil tensions propagate relatively rapidly through the radon barrier, equilibrating to steady state conditions within a few years. By applying a cyclical upper boundary condition based on measured monthly average tensions in the upper portion of the radon barrier, relative soil tensions in the radon barrier were simulated for a period of 100 years. The modeling indicated that soil tensions in the radon barrier are currently at or near equilibrium, and that the radon barrier will remain unsaturated with time. The long-term percent saturation of the Shiprock radon barrier is predicted to be slightly less than the average 83.6 percent saturation measured in analyses of core samples in 1988.The modeling also showed that if the filter layer were to remain saturated year-round, then the saturated moisture front would propagate downward through the entire radon barrier within a year. However, saturation of the radon barrier in the future is unlikely, as its low hydraulic conductivity limits downward migration of water, and evaporation removes excess water from the filter layer.

Based on data from the field study at Shiprock and the unsaturated flow modeling, it was concluded that the operating hydraulic conductivity of the Shiprock radon barrier is approximately 10^{-8} centimeter per second, and moisture conditions within the radon barrier are approaching a state of dynamic equilibrium. Radon barriers of similar UMTRA Project disposal cells in similar climates also are likely to remain unsaturated if potential evaporation from the filter layer exceeds precipitation for most of the year.

Basis of Performance of UMTRA Disposal Cells

It has been recognized that the properties and function of UMTRA rock covers will undergo changes with time. These changes may include soil genesis with the radon barrier, silting of the rock layers, and plant succession. Although these changes may occur, they do not substantially influence the performance measures that include reducing radon emanation and meeting groundwater standards at the POC. Even if the radon barrier increases in saturated conductivity with time due to natural processes, the low conductivity of tailings materials underneath the cover behave as an infiltration barrier to hold any infiltrated moisture until it can be evaporated. It is probable that the rock covers may eventually perform satisfactorily as ET covers. In addition, in arid to semi arid regions in the west, tailings will not resaturate after transient drainage of entrained moisture in the tailings has occurred. It is important that the cover remains graded so that water will not pond and increase infiltration by focused recharge.

In addition, the performance of an UMTRA disposal cell is not just dependent on low infiltration through the cover, it also relies on minimal transmission of infiltration vertically through the tailings, geochemistry of subpile soils to attenuate chemical constituents, and dilution from groundwater underflow. The slimes and transitional tailings have a similar low unsaturated conductivity to the cover. In some cases, groundwater beneath the disposal cell is Class III (limited use) and supplemental standards apply so that the cover design only needs to be ALARA. Biointrusion of plants into tailings is limited by the concentration of salts in the tailings that would create either an osmotic stress or sodium hazard for plants.

Monitoring Performance of the Disposal Cell

Although not required by regulation, an easy and cost-effective way to test the continued efficacy of rock covers is to core the radon barrier and tailings periodically and analyze moisture contents from approximately the same locations. Records were kept in the UMTRA Project files on moisture contents of radon barriers at the time of placement and moisture contents of tailings when they were relocated. As transient drainage of tailings occurs and the covers evolve to adjust to the arid climates, moisture content should stay relatively constant or decrease gradually with time. In cases where slime and transitional tailings are instrumented, meaningfu change in moisture content will be slow. In some cases, the change may be below the sensitivity of the instruments, beyond the time of instrument function in a highly corrosive geochemical environment, or past the period of project funding. Vertical movement of water through slime and transitional tailings materials once they become unsaturated is negligible.

References

DOE 1989. *Moisture Contents and Unsaturated Conditions,* UMTRA Project Radon Barriers, U.S. Department of Energy, UMTRA Project Office, Albuquerque Operations Office, Albuquerque, New Mexico

DOE 1991. *Analysis of infiltration though a Clay Radon Barrier at an UMTRA Disposal Cell,* UMTRA Project Radon Barriers, U.S. Department of Energy, UMTRA Project Office, Albuquerque Operations Office, Albuquerque, New Mexico

Note: These papers are available on the conference website:
http://www.nrc.gov/about-nrc/regulatcry/decommissioning/public-meetings/materials2010.html

SESSION 6:

RECOMMENDATIONS ON ASSESSING ENGINEERED BARRIER PERFORMANCE, IDENTIFYING FUTURE RESEARCH NEEDS, AND DISCUSSING EXISTING GUIDANCE

Session Chairs:
Thomas Nicholson, NRC/RES and
Hans Arlt, NRC/FSME

Technical Reporters:
Mark Fuhrmann, NRC/RES and
Allen Gross, NRC/FSME

8.1   Overview and Insights

Over the last few years, research findings from various governmental, regulatory and academic organizations, including NRC/RES, have raised technical questions regarding performance of engineered surface barriers and liners, and assumptions in their performance assessments. Operators, regulators, and researchers of engineered barriers recognize the need to discuss these technical issues for assuring long-term performance to protect human health and the environment. In April 2009, FSME requested a briefing by RES staff on engineered covers to the States of South Carolina and Washington, and Energy Solutions (Barnwell Site operator). Based upon discussions at this meeting, Chris McKenney, Chief, Performance Assessment Branch, FSME and Tom Nicholson, Senior Technical Advisor, RES/DRA suggested that a technical workshop be organized to further inform the NRC licensing staff and States on lessons learned on the state-of-the-practice in evaluating the performance of engineered systems to isolate radioactive waste. Other drivers for organizing the workshop were technologies and designs for the new Low-Level Radioactive Waste (LLW) facility being licensed in Texas (western Andrews County - Waste Control Specialists) and research sponsored by DOE and EPA on evaluating long-term performance and maintenance of engineered covers.

Research sponsored by NRC's Office of Research (NRC/RES), and presented by Professor Craig Benson at the workshop, will be documented in detail in a NRC contractor report (NUREG/CR- 7028 *in press*) entitled "Engineered Covers for Waste Containment: Changes in Engineering Properties and Implications for Performance Assessment." This report's findings and conclusions pertain to changes in both saturated and unsaturated soil hydraulic properties due to degradational processes (e.g., wetting and drying, freezing and thawing, and soil development). These processes are relevant to all engineered soil layers with particular importance in compacted clay liners. Previous field studies published and documented have shown similar findings. Frequently, aggregates and planes of weakness or macropore structures may continue to exist after cover construction, and environmental processes related to pedogenesis may cause degradation related to in situ soil development. These topics and others were highlighted in the workshop agenda, and discussed in the session presentations and panel discussions. In addition, the workshop provided detailed information on field characterization and monitoring tools and methods, and numerical models to analyze short- and long-term performance of engineered covers used at uranium mill tailings, decommissioned sites, and LLW facilities.

Two of the most consistent themes repeatedly mentioned by the workshop participants were: 1) the need to more actively monitor and quantify engineered system behavior so as to know the record of performance and better understand the key processes effecting performance; and (2) the need for better communication and exchange between the specialists of various fields, (e.g., engineers, pedalogists, soil scientists, ecologists, modelers, etc.) involved with the design, construction, and maintenance of engineered barriers. These insights were supported by technical presentation details and discussions by all engaged participants, and were repeated throughout the three-day workshop.

8.2   Observations

The observations, insights, and recommendations documented below were made by participants during the workshop proceedings.

8.2.1 Principal Observations

- More effort is needed to facilitate communication between Federal and State regulators, facility operators and the technical community in the following topical areas: degradation processes and changing performance of engineered barriers thru time; monitoring (short-term) programs and data analysis; model support (long-term) to strengthen the technical basis of the estimated parameter values and their time-dependent or process-based changes; modeling of processes within the barriers, especially engineered surface covers; discussion of lessons learned and practical examples of performance failures and successes based on field observations.

- General consensus that groups involved with engineered barrier performance need to communicate more frequently, provide access to systematized databases; and coordinate their efforts to raise the general state-of-the-knowledge in engineered system performance.

- The performance of cover systems evolve toward an equilibrated state more rapidly than originally anticipated (with the exception of geomembranes where the longevity of these components may have been conservatively estimated in earlier research). The alteration or evolution of covers can lead to degradation of a resistive cover to an evapotranspiration cover. The level of monitoring and model support should be risk informed and performance based.

- General alterations in certain cover properties were observed indicating variations from the designed as-built properties. These alterations were due to exposure to natural environmental conditions which caused root intrusion and soil development resulting in changes to saturated to unsaturated soil properties. These alterations may cause preferential flow in the low-permeability radon barriers and alterations to soil water storage properties.

- It is imperative to understand the conceptual model of the engineered cover and its components including water budget analyses for each system component over time, and each component's risk significance within the total system.

- One important conceptual approach for understanding engineered system performance is to assess the "energy" of that system and how much energy is required to maintain the engineered system over time. The further away the system is from natural equilibrium, the more energy will be required to maintain its functions. An energy balance should be calculated using an approach similar to water balance assessments.

- Evapotranspiration or water-balance covers should take advantage of the pedogenesis processes. For example, if soil properties change due to pedogenesis, then the establishment of flora may increase or maintain overall performance through greater root densities and water uptake due to the soil property changes.

- There appeared to be no urgent problems identified with engineered surface covers associated with facilities governed by the Uranium Mill Tailings Radiation Control Act of 1978 (UMTRCA) as shown by limited air and water quality monitoring. However, there was general consensus that uranium mill tailings sites need more active monitoring in the covers, tailings, and below the tailings in keeping with a system approach.

8 - 3

- There was general consensus that a total systems approach to monitoring and modeling should be pursued. Engineered covers and liners should not be looked at in isolation from the entire waste disposal system. A hierarchical and iterative approach to identify significant processes and components should be used to understand where further detailed modeling and/or monitoring is warranted.

8.2.2 Additional General Observations

- Continue to use a strategy of defense-in-depth for the design of engineered systems. For example, uranium mill tailings disposal sites often use layered tailings with the less radioactive material overlying the more radioactive material; a soil/frost protection layer in addition to radon barrier; and the incorporation of soil materials into the rock covers to provide additional radon protection.

- Institutional controls have been ineffective at some sites where fences and other restrictive structures and means were not effective in precluding human access to the site. For example, impediments constructed to prevent public access to uranium mill tailings disposal sites were removed and sites were accessed by members of the public.

- Recommendations were made by some participants to reconsider the time period of institutional control for LLW disposal sites. For example, the change could be a more adaptive and site-specific approach consistent with UMTRCA sites and provide for more flexible responses and activities. This change may necessitate dedicated funding sources for long-term control which should include monitoring.

- Newer guidance is needed on the design of covers and liners, their associated construction techniques and monitoring that incorporate new knowledge and techniques. Guidance should be flexible allowing for improvements in knowledge, experience, and techniques over time.

- A common data repository is needed for current and future information on engineered barrier performance. This proposed data repository could be a multi-agency effort with ongoing maintenance.

8.2.3 Specific Observations

- Many specific recommendations were made during the workshop with regard to monitoring significant components or processes of the barrier system. The frequency and location of monitoring devices needs to be coupled to model simulation results.

- "Bands of armor" can be incorporated in the design of covers using the results of landform evolution modeling to determine the probabilities and locations of gully formation.

- Modular systems of engineered barriers, as opposed to a large monolithic design, may encourage ground-water recharge between modules and thereby increase dispersive mixing and dilution.

- Recommendation to use geophysical investigations with caution on their limited usefulness, both spatially and temporally, of focusing on point sensors and sampling.

- Recommendation to design for catastrophic events. For example, site disposal facilities or ridge tops instead of in valley bottoms, if investigations have shown catastrophic scouring of the valley bottoms in the past.

8.2.4 Monitoring and Modeling Observations

- Monitoring and modeling is an iterative process. Modeling can focus monitoring by identifying key processes and parameters or disconnects between field observations and model results. Similarly, the results of monitoring provide feedback to refine models and improve the understanding of the system. Monitoring should be driven by the purpose of the project. The technology exists to monitor components of the system important to safety and environmental concerns. However the cost of monitoring may be limiting.

- The specificity of the engineered design should be commensurate with the level of risk, the anticipated costs of monitoring, and the cost of corrective actions to remedy failures or under-performance of the total system.

- In modeling the total system performance, it is important to clearly assess the designed function of each cover component. If the interrelationships among these components are not understood, it may lead to unintended consequences of performance.

- Monitor safety-significant components of the facility including internal containment system monitoring to enhance performance. Monitor the containment structure to understand processes and to identify precursors of problems. This approach is more effective than perimeter compliance wells that do not provide an understanding of the processes that impact compliance.

- Monitor ecological/plant processes if they are potentially critical to cover performance.

- Monitor the unsaturated zone within and below the emplaced waste to the regional water table.

- Place monitoring devices within the engineered cover elements to obtain data on water infiltration and gaseous releases.

- Identify methods and guidance for implementing large-scale lysimeter monitoring to quantify preferential fluxes through the cover to the emplaced waste, and through to the underlying unsaturated zone and deeper saturated zone below.

- If the engineered barrier performance is important and associated with significant uncertainty, monitor and confirm ground-water releases, as well as radon emissions for sites with designed-based compliance.

- Use remote sensing to assess landscape-scale environmental conditions that are indicative of facility performance.

- Differential settlement and human intrusion could be potential significant failure modes for cover performance.

- Monitor differential settlement and human intrusion and their potential to dramatically alter the cover performance.

- Monitor microbial activity that may affect drainage layers (e.g., biofouling) and geosynthetic performance.

- More information is needed on biotic activity and its relation to the performance of engineered surface covers (i.e., information such as root and burrow penetration depth distribution and the dynamics of ecological succession).

- Simulation codes need to better incorporate ecological succession and climatological changes. They need to account for episodic rates, as well as dynamic disturbances such as fires and changes in biota. All model assumptions and evaluations should be made available for assessment.

- Risk significant information should be incorporated in the overall total systems performance model. The engineered barrier degradation processes should also be integrated into the model.

- Monitor and model the ecosystem with an emphasis on quantifying baseline conditions to model anticipated future human and natural events that may significantly affect performance.

- Identify and test assumptions in performance assessments for past uranium Title I sites, present uranium Title II sites, and future engineered barrier systems.

- Coupling of hydrology, erosion, and plant succession should be considered in the short- and long-term performance assessment of the engineered systems.

- The coupling of monitoring and modeling activities should focus on performance indicators that can be both monitored and modeled.

- In the prediction of both short- and long-term performance of the engineered system components and its surrounding environments, uncertainty assessments are required.

- Assess conceptual, parameter, and scenario uncertainties with emphasis on alternative conceptual models that include significant features, events and processes (FEPs) that can affect long-term performance.

- When considering future scenarios, a screening approach or framework that includes natural analogues would be helpful. For example, the pedogenesis of a soil class exposed to local climatological conditions and ecological changes would constitute a particular scenario. A range of possible alternative scenarios using paleoclimate conditions should then be considered. These scenarios should also identify degradation processes affecting performance, e.g., different barrier types for different types of ecologic and climate states.

8.2.5 Model Support Observations

- Continue to improve model support and confidence building activities. This process should continue after the initial evaluation. Re-evaluations should be performed at least every 5 years to confirm that design and construction matches expected engineered system performance.

- Develop and implement strategies to obtain and evaluate information needed to support both short- and long-term modeling results.

- Model support should be risk informed and needs to evolve with different stages of the project and should be specific to each site's characteristics. For example, new material may be introduced in the actual construction. This information needs to be communicated to the modeler for re-assessment of the assumptions and scenarios being simulated.

- Improve communication between model developers and those who provide model support. These communications should be documented in the quality assurance/quality control (QA/QC) program.

- Develop a "Catalogue of Natural Analogs" for different climatic and environmental settings throughout the United States. This catalog of analogs could provide insight into the ecology, climate, and erosive forces for long-time periods. It could provide bounding scenarios of reasonable future states. For example, pedogenic carbonates provide an analog of a capillary barrier on the order of 10,000 years. Analogs of ecological succession can also provide clues about possible future changes in the ecology and evapotranspiration on engineered covers. However, some types of analogs must be used with caution, since unknown analogs may not have survived into the present, and past environmental exposure may not be known.

- Soil development will be a function of climate, parent material, topography, biology, and time. Pedogenic features will affect hydraulic processes and plant communities at various scales, both spatially and temporally. The modern landscape provides an analog or record of soil evolution pathways that can support long-term assessments of covers.

- The important properties of clay barriers quickly degrade when exposed to the surface environment (i.e., wet-dry and freeze-thaw cycles). By locating the clay barriers sufficiently below an overlying cover (e.g., soil) unit, certain degradation processes may be minimized. More information is needed on the relationship between the depth of the soil (clay) barrier and the thickness of the overlying cover material that may affect the rate of degradation or property changes.

- Understanding the geochemical and microbial degradation processes and conditions are fundamental to designing the cover appropriately.

8.2.6 Processes and Performance Observations

- Performance of current geosynthetics appears promising (i.e., potential service life of hundreds of years). However, a paucity of relevant data exists relating to the effect of low-level radioactive waste (LLW) on service life of geomembranes. Impact of radiation is determined by total absorbed dose in the presence or absence of oxygen. If oxygen is available, irradiation can provide the activation energy for oxidation to occur. Service life could be reduced by heat generation, high pH leachate, or irradiation from the waste itself.

- The greatest short-term risk to geomembranes is caused by membrane punctures resulting from construction activities, animal burrowing, and excessive differential settlement. Typical modes of degradation of geomembranes include oxidation, extraction (e.g., diffusion of antioxidants into the surrounding environmental), biological degradation, UV, and thermal degradation. The most significant short- and long-term ecological degradation processes are root penetration and growth, plant decomposition, animal intrusion, and bioturbation of soils.

- QA/QC guidance and confirmatory testing of geosynthetic materials and their installation need further development. Since geomembranes are relatively fragile materials in hostile environments, strict QA/QC is critical. Quality construction of cover is important and should be documented in the QA/QC program. The QA/QC pedigree must be robust.

- Erosion/deposition rates may vary spatially and temporally across a cover. Long-term erosion often concentrates in gullies which do not uniformly erode over their entire length. Peak erosion depth may translate into a total breach of a portion of the cover and possible failure.

Although the current state of the landform evolution model software has many drawbacks, it can assist in predicting where the most sensitive parts of the cover are located. Next steps include coupling ecology succession with soil development and landform evolution.

8.3 NRC Staff Observations and Recommendations

NRC staff reviewed and studied the usefulness and applicability of most of these recommendations and insights from the workshop. NRC staff agrees with many of these recommendations and makes the following specific observations and comments:

- Engineered barriers have only been in use for a few decades to isolate hazardous and radioactive waste from human populations. It is therefore not surprising that research and documentation continues on the overall assessment of the performance of engineered barrier systems for various lengths of time and for very diverse sites. Therefore, individuals and groups involved with engineered barrier performance need to communicate more frequently and coordinate their efforts.

- As-built properties for each component of an engineered barrier cannot be expected to remain indefinitely constant when exposed to the forces and processes of the environment. Designed-based criteria for such features are not sufficient. Actual performance needs to be monitored and modeling predictions need to be validated by supporting data and analyses.

- Until the interrelationship between these processes and barrier performance is better understood, more data should be collected and analyzed. The level of monitoring and data collection should be risk-informed and commensurate with the level of performance required and expected. Data collected should be relevant to the time period of reliance (performance) on the engineered system barrier. If performance is needed for relatively long-time periods [e.g., engineered surface covers designed for waste-incidental-to-reprocessing (WIR)], then data collection would help confirm the main assumptions and provide a more robust and site-specific model support. Based on the information presented during the workshop, NRC staff would expect increased data from the following significant areas:

- o Increased interest in the long-term application of evapotranspiration and geomembranes to minimize water infiltration through disposed waste.

- o Increased interest in pedogenic processes, biotic activities, and bioturbation occurring within the cover soils and drainage layers.

- A heavily monitored test pilot cover, or a reserved section of the engineered barrier for detailed monitoring purposes, could provide valuable data on: characterizing changes in the properties of cover system components; developing in-situ methods to detect such changes; and developing and validating predictive methods for performance assessments that account for time-dependent engineering properties.

- The regulations in 10 CFR Part 61 cover all near-surface LLW disposal activities from site selection through facility design, licensing, operations, closure, and postclosure stabilization to the time period when active institutional controls end. The regulation requires the use of engineered features in concert with the natural characteristics of the disposal site to contain and isolate the LLW. It also includes long-term numerical performance objectives and other technical criteria.

- In addition to licensing LLW near-surface disposal facilities, NRC also regulates uranium mill tailings under the Uranium Mill Tailings Radiation Control Act of 1978 (UMTRCA). This legislation addresses two types of sites: (1) those that were inactive or no longer being used when UMTRCA was passed, referred to as Title I sites; and (2) sites that were active or issued a license after UMTRCA was passed; referred to as Title II sites. Title I sites are regulated under the requirements of 40 CFR Part 192 and have a general license under 10 CFR 40.27. Currently, the standards are being revised by the U.S. EPA. The NRC staff discussed with EPA staff potential revisions to these standards to be more risk-informed, performance-based rather than prescriptive designed-based which was discussed at the workshop. Such changes to the standards would in turn encourage NRC to consider revisions to its own regulations in Appendix A to 10 CFR Part 40. These revisions could include long-term numerical performance objectives, and place a greater emphasis on a performance monitoring strategy beginning with the barrier design. Guidance could be revised to allow for greater flexibility with regards to changing knowledge, experience, and techniques over time.

- NRC licensing staff agrees with a total systems approach to monitoring and modeling. Engineered covers and liners should not be assessed in isolation from the entire waste disposal system, rather they should be incorporated into a graded and iterative approach for identifying significant processes and components requiring further detailed modeling and/or monitoring. This approach is in alignment with NRC's general defense-in-depth approach for performance.

- In line with the recommendations made at the workshop, NRC licensing staff has contracted with the Southwest Research Institute's CNWRA to compile and develop a database of site-specific data on radon and groundwater, evaluate site-specific engineered cover performance, and recommend future site-specific radon and groundwater monitoring and data analyses for all Title II-in-closure sites. The CNWRA contract might result in additional recommended long-term work based on the analyses of monitoring data or the recognition of missing information that is needed at these sites.

- NRC staff is also aware of U.S. Department of Energy's Legacy Management (LM) Cover Renovation Research, or RECAP program, and will follow the findings and results of this program. This attempt to transform conventional covers into ET covers, reduce soil bulk density (compaction), increase soil water storage capacity, and enhance establishment of favorable vegetation will provide valuable input when deciding which future directions to follow in engineered performance of the cover and design.

- NRC staff will work and cooperate with other Federal and State agencies that have a role in determining or maintaining performance for engineered surface barriers. For example by setting up a subcommittee with other such agencies within the framework of the Interagency Steering Committee on Radiation Standards (ISCOR). Currently, NRC is part of the DOE Landfill Partnership. This partnership consists of members of the Consortium for Risk Evaluation and Stakeholder Participation (CRESP) which includes regulatory stakeholders from EPA, NRC, DOE, and various State agencies. These partners will be asked to identify applied research activities necessary to resolve technical issues and develop or recommend technical approaches to remedy technical inconsistencies in existing regulations. The partnership was initiated in FY2010 as a vehicle to address these issues, and will conduct applied research and facilitate technical dialogue needed to build confidence in technologies used for on-site disposal facilities, the methodology used to design and assess these facilities, and the systems used for monitoring long-term performance.

APPENDIX A: Workshop Agenda

Workshop on Engineered Barrier Performance
Related to Low-Level Radioactive Waste,
Decommissioning and Uranium Mill Tailings Facilities

Time: August 3 - 5, 2010, 8:30 am – 6:00 pm (EDT)
Location: NRC Headquarters Auditorium, 11555 Rockville Pike, Rockville, MD 20852.
WebStreaming: http://video.nrc.gov/live/

August 3, 2010 (8:30 am – 12:30 pm EDT) Tuesday

Session 1: Introductions and Orientation

8:30 am

Welcome and Introductions
James Lyons, Deputy Director, NRC/RES (5 min.)
Larry Camper, Director, NRC/FSME/DWMEP (5 min.)

8:40 am

Discussion of Workshop Objectives, Goals and Agenda
Hans Arlt, NRC/FSME/DWMEP (10 min.)

Objectives: Facilitate communication of Federal agencies' research and State regulatory experiences on the workshop topics to the technical community, and to discuss degradation processes and changing performance of engineered barriers, monitoring (short-term), model support (long-term), and modeling of processes within the barriers, especially engineered surface covers. Discuss lessons learned and practical examples of performance failures and successes based on field observations. Share information on research results, existing guidance, and identify potential improvements to guidance.

Goals: Identify lessons learned and recommendations to maintain adequate engineered barrier performance; to include areas for future research, and to identify potential needs for modifying and updating guidance.

8:50 am

Identification and Differentiation of Engineered Barrier Types
by Function and Design
Professor Craig Benson, University of Wisconsin/CRESP (20 min.)

- **Surface Covers** – conventional covers with clay or composite (clay-geomembrane) barriers; water balance covers that control percolation by balancing soil water storage and water removal via evapotranspiration.

A - 1

- **Bottom Liners** – subsurface barriers along the base and sidewalls of disposal facilities constructed with clay barriers, geomembranes, geosynthetic clay liners, and combinations thereof.
- **Cover and Bottom Liner Functions**: control water percolation into waste; control gases and radon release; maintain stabilization/prevent erosion; deter inadvertent intruders; and minimize contaminant transport or a combination thereof.

9:10 am

Overview of Engineered Barrier Performance and
Regulatory Compliance Criteria
Jacob Philip, NRC/RES and David Esh, NRC/FSME (15 min.)

- Experiences with different engineered barrier types and their various components (e.g., DOE UMTRA sites to include Title I and II sites; LLRW facilities; WIR multi-layer covers; ACAP examples)
- NRC guidelines on engineered barrier performance, or on monitoring disposal sites of various waste types can be found in NUREG-1757 for complex materials decommissioning, NUREG-1854 for WIR, NUREG-1620 for radioactive mill tailings, and NUREG-1388 for LLW. NUREG-1623 presents methods, guidelines, and procedures for designing erosion protection, for long term stabilization
- NRC experiences and timeline
- Recent research and publications

9:25 am

Experience of the States in Regulating Facilities
Involving Engineered Covers and Liners
Session Chairs:
Stephen Salomon, NRC/FSME and
Susan Jablonski, TCEQ, State of Texas (80 min.)
Technical Reporter:
Douglas Mandeville, NRC/FSME

- Overview of research activities and findings with emphasis on practical insights on monitoring, modeling and confirming short- and long-term performance of engineered systems

Questions for Presenters:

- What are your State's regulatory activities and findings which confirm short- and long-term performance of engineered systems with emphasis on practical insights on monitoring and modeling?

- What are your siting regulations regarding engineered barriers (1) degradation processes that change performance; (2) monitoring devices and systems; (3) codes and modeling experiences; and (4) model support to gain confidence in long-term performance?

- How do you see these regulations evolving based upon experiences?

Presentations

9:25 – 9:35 am	**Modeling and Monitoring of Barrier Performance for the Planned Texas Low-Level Radioactive Waste Disposal Facility** Susan Jablonski, P.E., Peter Lodde, P.E., and Abel Porras, P.E. Texas Commission on Environmental Quality (TCEQ), State of Texas
9:35 – 9:45 am	**Utah Clive Low-Level Radioactive Waste Facility** Loren Morton, Utah Division of Radiation Control, State of Utah
9:45 – 9:55 am	**Overview of the Performance and Use of Engineered Barriers at the Barnwell LLRW Disposal Site** Susan E. Jenkins, Division of Waste Management, South Carolina Department of Health and Environmental Control (SC DHEC), State of South Carolina
9:55 – 10:05 am	**Washington State's Experience with Decommissioning and Evaluation of Cover Designs for Low-Level Radioactive Waste and Uranium Mill Tailings Facilities** Gary Robertson, Office of Radiation Protection, Washington State Department of Health, State of Washington
10:05 – 10:15 am	**Colorado Experience with Waste Repository Covers and Caps** Lawrence J. Bruskin, P.E., Senior Engineer, and Steve Tarlton, P.E., Radiation Program Manager, Colorado Department of Public Health & Environment, Hazardous Materials and Waste Management Division, State of Colorado

Panel Discussion by Presenters and Panelists (30 min.)

10:15 – 10:45 am   **Panelists:**

Steve Austin, Hydrologist for the Navajo Nation UMTRA sites, Navajo Environmental Protection Agency, Navajo Nation

Robert Paneuf, Acting Director, Bureau of Hazardous Waste & Radiation Management, Division of Solid & Hazardous Materials, Department of Environmental Conservation, West Valley LLW Facility in the State of New York

10:45 am   **BREAK** (15 min.)

11:00 am

Federal Agencies and DOE National Laboratories
Session Chairs:
Jacob Philip, NRC/RES and
Brian Andraski, U.S. Geological Survey (90 min.)
Technical Reporter:
George Alexander, NRC/FSME

- Overview of research activities and findings with emphasis on practical insights on monitoring, modeling and confirming short- and long-term performance of engineered systems

Questions for Presenters:

- What performance assessment (PA) was done to predict dose due to gaseous and fluid releases from the facility?

- What laboratory and field tests were performed to obtain input parameters for the PA modeling?

- What field and laboratory tests were performed, and what measurements were taken to validate PA model results?

- Is field monitoring continuing and at what intervals, to validate that the facility is continuing to perform to regulatory criteria?

- What maintenance and repair activities are conducted to remediate the facility if regulatory criteria are not being met?

- Are the PA's that were conducted for the sites and the laboratory/field test results publically available?

Presentations

11:00 – 11:02 am	**Introduction** Jacob Philip, NRC/RES and Brian Andraski, USGS
11:02 – 11:12 am	**USACE Experience with HTW Containment Systems** Kevin Pavlik, U.S. Army Corps of Engineers
11:12 – 11:22 am	**The Legacy Management UMTRCA Program** Richard Bush, DOE/Legacy Management (DOE/LM)
11:22 – 11:32 am	**EPA's Review of Its Regulatory Requirements for Uranium and Thorium Mill Tailings: 40 CFR Part 192** Loren Setlow, U.S. EPA

11:32 – 11:42 am	**Investigations Supporting Performance Verification of Engineered Barrier Systems**
	Joel Hubbell, Idaho National Laboratory

11:42 – 11:52 am	**Savannah River Site E-Area Low-Level Waste Facility Subsidence Studies**
	Mark Phifer, Savannah River National Laboratory

11:52 am – 12:02 pm	***DOE/EM Overview***
	Roger Seitz, Savannah River National Laboratories for Martin Letourneau, DOE/Environmental Management (DOE/EM)

Panel Discussion by Presenters and Panelists (28 min.)

12:02 – 12:30 pm **Panelist:**

David W. Esh, U.S. NRC

12:30 pm

LUNCH (60 min.)

August 3, 2010 (1:30 – 5:30 pm EDT) Tuesday

Session 2: **Degradation Processes and Performance Evolution of Engineered Barriers**
Session Chairs:
Craig Benson, University of Wisconsin/CRESP and
W. Jody Waugh, S.M. Stoller LLC
Technical Reporter:
Brooke Traynham, NRC/FSME

Topics to be Considered:

- Degradation processes affecting barrier components (e.g., geomembranes, GCLs, drainage layers)
- Climatic factors contributing to degradation in the near term and long term
- Environmental equilibrium: plant succession, climatic variability, and geomorphic processes due to changes in local hydrology
- Anthropogenic impacts on engineered barriers in covers
- Impacts of erosion
- Microbial processes that affect barrier materials and drains (biofouling)
- Geochemical processes that affect degradation of barriers and drains (chemical erosion, embrittlement, and clogging of drainage)

Questions for Presenters:

- For all types of covers, what are the most significant short-term and long-term degradation processes causing increases in radon release, water percolation, erosion, and bio-uptake?

- For all types of liners, what are the most significant short-term and long-term degradation processes causing increased water and contaminant flux?

- How will climatological and ecological changes affect degradation processes (e.g., at humid, temperate sites, as well as for dry, cold sites)?

- How can degradation processes be minimized, and radon release, percolation, erosion, and bio-uptake be reduced for various ecologies and climates (e.g., QA/QC, installation, type of cover, material, etc.)?

- Can the desired changes to reduce one process cause the undesired increase of another; for example, activities that reduce erosion inadvertently cause an increase in water percolation? How can such unintended consequences be avoided?

- How can our understanding of degradation processes be used to improve the designs and performance of covers and liners?

Presentations

1:30 – 2:00 pm	**UMTRA Experience Monitoring Degradation Processes and Their Effects on the Performance of Covers** Jody Waugh, SM Stoller Corporation (DOE/LM), Grand Junction, CO
2:00 – 2:30 pm	**Soil Development Processes and Their Effects on the Performance of Covers** Craig Benson, Geological Engineering, University of Wisconsin
2:30 pm	**BREAK** (10 min.)
2:40 – 3:10 pm	**Geomorphological and Landform Processes and Changes in the Performance of Covers** Gary Willgoose, Australian Professorial Fellow in Environmental Engineering, University of Newcastle, Callaghan, Australia
3:10 – 3:40 pm	**Ecological Processes and Changes in the Performance of Covers** Steven Link, Department of Science and Engineering, Confederated Tribes of the Umatilla Indian Reservation, Pendleton, OR
3:40 – 4:10 pm	**Degradation Processes and Changes in the Performance of Geosynthetics**

Kerry Rowe, Vice-Principal and Professor of Civil Engineering, Queen's University, Kingston, Ontario

4:10 pm   **BREAK**   (10 min.)

Panel Discussion by Presenters and Panelists   (70 min.)

4:20 – 5:30 pm   **Panelists:**

Bill Albright, Desert Research Institute/UNV

Bob Phaneuf, New York State Department of Environmental Conservation (NYS DEC)

Mark Phifer, Savannah River National Laboratory

Kevin Leary, DOE-Hanford

5:30 pm   **Opportunity for Public Questions and Comments**

Please call-in to: **1-888-566-6344**   Passcode: **15103**

6:00 pm   **ADJOURN**

August 4, 2010 (8:30 am – 12:30 pm EDT) Wednesday

Session 3: **Experience with Monitoring Devices and Systems Used to Measure Performance**
Session Chairs:
William Albright, Desert Research Institute/UNV and
Craig Benson, University of Wisconsin/CRESP
Technical Reporter:
Robert Johnson, NRC/FSME

Topics to be Considered:

- Monitoring of short-term performance processes and indicators of percolation, leakage, and radon flux
- Monitoring of long-term performance processes and indicators using indirect (time-lapse imagery or geophysical surveys) and direct monitoring (large-scale pan lysimeters)
- Remote sensing and surveillance
- Direct measurement of percolation rates and radon fluxes over specified intervals
- Meteorological monitoring of rainfall, snow cover, temperature, and evapotranspiration
- Leachate collection and analysis for liners
- Sampling of contaminants and soil water chemistry to detect failure modes

- Monitoring of degradation processes on, and within, the barrier that modify the barrier from "*as built*" performance metrics to a longer-term performance level
- Monitoring to verify assumptions in PAs and modeling predictions
- Remote monitoring methods

Questions for Presenters:

- What areas should be monitored for significant degradation/performance (i.e., what are the important process and components)?

- Which barrier systems can be effectively monitored (*in situ* and remotely), and for how long?

- What tools, techniques, and methodologies are available for monitoring, and where/when should they be applied?

- What type and level of monitoring should be done (data sufficiency), and for how long?

- Does monitoring in the short-term provide insights and possible understanding of long-term issues?

- How important are information gaps in monitoring?

Presentations

8:30 – 8:55 am	**In Search of the Perfect Cap: 15 Years of Performance Data from the Prototype Hanford Barrier** Andy Ward, Pacific Northwest National Laboratory
8:55 – 9:20 am	**ACAP: Monitoring cover performance and changes in performance with drainage lysimeters, instruments, and exhumations** Bill Albright (DRI) and Craig Benson (UW)
9:20 – 9:45 am	**Monitoring Contaminant Strategies: Tools, Techniques, Methodologies and Modeling Approaches** Tim Gish, Audrey Gruber, Yakov Pachepsky, U.S. Department of Agriculture/Agricultural Research Service, Beltsville, MD
9:45 am	**BREAK** (10 min.)
9:55 – 10:20 am	**Aerial remote sensing as a component of closure cap monitoring** John Gladden, Savannah River National Laboratory
10:20 – 10:45 am	**Differential Settlement and its Importance on the Performance of Cover Systems at Radiological Waste Disposal Facilities** Bob Bachus, Geosyntec Consultants

10:45 am   **BREAK** (15 min.)

Panel Discussion by Presenters and Panelists (90 min.)

11:00 – 12:30 pm   **Panelists:**

Brian Andraski, US Geological Survey

Bill Kustas, USDA/ARS

12:30 pm   **LUNCH** (60 min.)

August 4, 2010 (1:30 – 5:30 pm EDT) Wednesday

Session 4: **Modeling Experiences in Performance Assessment and Evaluation of Performance Monitoring**
Session Chairs:
David Esh, NRC/FSME and
Thomas Nicholson, NRC/RES
Technical Reporter:
Christopher Grossman, NRC/FSME

Topics to be Considered:

- Water balance models to evaluate storage capacity, infiltration and deep percolation
- Assess environmental conditions
- Assess failure modes and changes to materials and system components over time
- Small- (point) versus large-scale (average) estimates of flux and perturbations
- Estimate percolation rates through covers at different scales
- Estimate radon flux through various covers (especially clay covers) over time
- Estimate long-term environmental equilibrium conditions related to natural and anthropogenic changes
- Issues of spatial/temporal scale and corresponding field-scale observations
- Time periods for evaluation (i.e., 0 – 5 years, 5 – 10 years, 10 – 50 years, 50 – 100 years, 100 – 500 years, 500 – 1,000 years, and greater than 1,000 years)

Questions for Presenters:

- When should numerical modeling of engineered barriers be performed?

- Over what time periods should performance simulations be considered?

- What are the criteria to determine the detail of modeling needed, e.g., should the actual processes changing a GCL be modeled?

- Which hydrologic, erosion, and mass wasting codes are recommended to better evaluate long-term performance of covers?

- What codes are recommended for simulating ecological evolution?

- What codes are recommended for predicting physical and chemical changes in soil properties and geosynthetic materials?

- How should ecological and climatological changes be incorporated into performance simulations?

- What input data and parameters are required for these codes and is this information available?

Presentations

1:30 – 1:55 pm	**Development of an Integrated Probabilistic Model of Radiological Fate and Transport in an Engineered Cover** John Tauxe, Neptune and Company
1:55 – 2:20 pm	**Practical Considerations for Modeling and Monitoring of Engineered Barriers Performance** Roger Seitz, Savannah River National Laboratories
2:20 – 2:45 pm	**Near-Term Hydrological Performance Modeling of Covers** Craig Benson, University of Wisconsin/CRESP
2:45 pm	**BREAK** (10 min.)
2:55 – 3:20 pm	***Prototype Hanford Barrier Modeling*** Andy Ward, Pacific Northwest National Laboratory
3:20 – 3:45 pm	**Effects of Plant Succession on the Functioning of Engineered Covers and Modeling of Long-Term Successional Impacts Using the EDYS Ecological Simulation Model** Terry McLendon, KS2 Ecological Services Specialists, LLC
3:45 – 4:10 pm	**Applications of thermal remote sensing for multi-scale monitoring of evapotranspiration** Bill Kustas and Martha Anderson, U.S. Department of Agriculture/Agricultural Research Service, Beltsville, MD
4:10 pm	**BREAK** (10 min.)
4:20 – 5:30 pm	**Panel Discussion by Presenters and Panelists** (80 min.)

Panelists:

Robert Holt, University of Mississippi

Ming Zhu, DOE - Environmental Monitoring

Gary Willgoose, University of Newcastle, Callaghan, Australia

5:30 pm   **Opportunity for Public Questions and Comments**

                Please call-in to: **1-888-566-6344** Passcode: **15103**

6:00 pm   **ADJOURN**

August 5, 2010 (8:30 am – 12:30 pm EDT) Thursday

Session 5: **Experience with Model Support and Multiple Lines of Evidence
to Gain Confidence in Long-Term Performance**
Session Chairs:
Hans Arlt, NRC/FSME and
George Alexander, NRC/FSME
Technical Reporter:
Brooke Traynham, NRC/FSME

Topics to be Considered:

- Types of model support strategies and multiple lines of evidence
- Field evidence and laboratory tests to build confidence in performance
- ACAP exhumation and process audits to identify failure modes
- Lessons Learned from uranium recovery experiences and monitoring programs
- Model support commensurate with the risk significance
- Evaluate plant succession and soil development affecting long-term performance
- Landform stability as analogs to engineered barriers
- Attributes and evolution of stable landforms
- Time periods for evaluation (i.e., 0 – 5 years, 5 – 10 years, 10 – 50 years, 50 – 100 years, 100 – 500 years, 500 – 1,000 years, and greater than 1,000 years)
- Development of a performance confirmation program
- Develop a Screening Framework
- Develop a Catalog of Analogs
- Reality checks and use of success criteria to build confidence in short- and long-term performance

Questions for Presenters:

- What information or "lines-of-evidence" is needed to have confidence that an engineered surface cover or bottom liner will perform as predicted for 100 years?

- What information or "lines-of-evidence" is needed to have confidence that an engineered surface cover or bottom liner will perform as predicted for 100's to 1000's of years as ecologic settings and climates change?

Presentations

8:30 – 8:55 am	**Overview of Model Support (for Engineered Barriers)** Dave W. Esh, NRC/FSME
8:55 – 9:20 am	**Activities that Support the Scientific Credibility of Radioactive Waste System Performance Models** Abraham Van Luik, Carlsbad Field Office, DOE-Environmental Management (DOE-EM)
9:20 – 9:45 am	**Geomembrane Performance in Landfill Cover Systems** George R. Koerner, Geosynthetic Institute (GSI)
9:45 am	**BREAK** (10 min.)
9:55 – 10:25 am	**A Role for Natural Analogs in the Design and Long-Term Performance Evaluation of Earthen Covers for Uranium Mill Tailings** William J. Waugh, S.M. Stoller Corporation
10:25 – 10:50 am	**Arid Soil Evolution and Pedologic Development: Process Considerations and Applications to Engineered Barrier Design** Todd G. Caldwell[1], Michael H. Young[2], and Eric V. McDonald[1] [1]Desert Research Institute, Division of Earth and Ecosystem Sciences [2]University of Texas, Bureau of Economic Geology
10:50 am	**BREAK** (15 min.)
11:05 – 12:30 pm	**Panel Discussion by Presenters and Panelists** (85 min.)

Panelists:

Mark Phifer, Savannah River National Laboratory

Kent Bostick, Professional Project Services, Inc. (Pro2Serve)

John Walton, Univ. of Texas – El Paso

Kerry Rowe, Civil Engineering, Queen's University

12:30 pm	**LUNCH** (60 min.)

August 5, 2010 (1:30 – 5:30 pm EDT) Thursday

Session 6: **Recommendations on Assessing Engineered Barrier Performance,
Identifying Future Research Needs, and
Discussing Existing Guidance**
Session Chairs:
*Thomas Nicholson, NRC/RES and
Hans Arlt, NRC/FSME*
Technical Reporters:
Mark Fuhrmann, NRC/RES and Allen Gross, NRC/FSME

Significant Insights and Recommendations from Session Presentations and Panel Discussions

1:30 – 1:42 pm	**States Overview** by Susan Jablonski, Stephen Salomon and Douglas Mandeville
1:42 – 1:54 pm	**Federal Overview** by Jake Philip, Brian Andranski and George Alexander
1:54 – 2:06 pm	**Degradation Processes** by Craig Benson, W. Jody Waugh and Brooke Traynham
2:06 – 2:18 pm	**Monitoring** by Bill Albright, Craig Benson and Robert Johnson
2:18 – 2:30 pm	**Modeling** by Dave Esh, Tom Nicholson and Christopher Grossman
2:30 – 2:42 pm	**Model Support** by Hans Arlt, George Alexander and Brooke Traynham
2:42 p.m.	**BREAK (18 min.)**

3:00 p.m. Group Discussion and Summary of Recommendations (115 min.)

Formulate recommendations on how to evaluate short- and long-term engineered barrier performance:

- Identify degradation processes affecting performance, e.g , different barrier types for different types of ecologic and climate states
    - o Identify strategies for monitoring and modeling these degradation processes
- To evaluate overall performance, recommend total system monitoring strategy
- To evaluate overall performance, recommend total system numerical modeling strategy
- To gain confidence in overall performance, recommend strategies to obtain information and evidence needed to support short- and long-term performance model results
- Highlight research opportunities to fill information gaps
- Identify potential improvements to existing guidance
- Recommend follow-up coordination among workshop participants

A - 13

4:55 p.m.   **Opportunity for Public Questions and Comments** (30 min.)

                    Please call-in to: **1-888-566-6344**   Passcode: **15103**

5:25 p.m.   **Action Items and Follow-Ups and Thanks to the Attendees and Speakers**
                    Tom Nicholson and Hans Arlt, Workshop Co-Chairs

5:30 p.m.   **ADJOURN**

_____

                              END

APPENDIX B: Workshop Participants

James Lyons, Deputy Director, NRC/RES
Larry Camper, Director, NRC/FSME/DWMEP
Hans Arlt, NRC/FSME/DWMEP
Craig Benson, University of Wisconsin/CRESP
Jacob Philip, NRC/RES
David Esh, NRC/FSME
Stephen Salomon, NRC/FSME
Susan Jablonski, TCEQ, State of Texas
Douglas Mandeville, NRC/FSME
Loren Morton, Utah Division of Radiation Control
Susan E. Jenkins, South Carolina DHEC
Gary Robertson, Washington State Department of Health
Lawrence J. Bruskin, Colorado Department of Public Health & Environment
Steve Austin, Navajo Environmental Protection Agency
Robert Paneuf, New York Department of Environmental Conservation
Brian Andraski, U.S. Geological Survey
George Alexander, NRC/FSME
Kevin Pavlik, U.S. Army Corps of Engineers
Richard Bush, DOE/Legacy Management (DOE/LM)
Loren Setlow, U.S. EPA
Joel Hubbell, Idaho National Laboratory
Mark Phifer, Savannah River National Laboratory
Roger Seitz, Savannah River National Laboratories
W. Jody Waugh, S.M. Stoller LLC
Brooke Traynham, NRC/FSME
Gary Willgoose, University of Newcastle, Callaghan, Australia
Steven Link, Confederated Tribes of the Umatilla Indian Reservation, Pendleton
Kerry Rowe, Queen's University, Kingston, Ontario
Kevin Leary, DOE-Hanford
William Albright, Desert Research Institute/UNV
Robert Johnson, NRC/FSME
Thomas Nicholson, NRC/RES
Christopher Grossman, NRC/FSME
John Tauxe, Neptune and Company
Andy Ward, Pacific Northwest National Laboratory
Terry McLendon, KS2 Ecological Services Specialists, LLC
Bill Kustas, U.S. Department of Agriculture/Agricultural Research Service
Ming Zhu, DOE/Environmental Management (DOE/EM)
Abraham Van Luik, Carlsbad Field Office, DOE/EM
George R. Koerner, Geosynthetic Institute (GSI)
Todd G. Caldwell, Desert Research Institute/UNV
Kent Bostick, Professional Project Services, Inc. (Pro2Serve)
John Walton, Univ. of Texas – El Paso
Mark Fuhrmann, NRC/RES
Allen Gross, NRC/FSME
Timothy Gish, U.S. Department of Agriculture/Agricultural Research Service
Robert Holt, University of Mississippi

APPENDIX C: Acronyms

ACAP	Alternative Cover Assessment Program
CCL	compacted clay liner
CERCLA	Comprehensive Environmental Response, Compensation, and Liability Act
CFR	Code of Federal Regulations
CNWRA	Center for Nuclear Waste Regulatory Analyses
CRESP	Consortium for Risk Evaluation and Stakeholder Participation
DOE	Department of Energy
DOE/EM	DOE's Office of Environmental Management
DOE/LM	DOE's Office of Legacy Management
DRI	Desert Research Institute
DVZ	Deep Vadose Zone
EBS	engineered barrier systems
EDYS	Ecological Dynamics Simulation
EPA	United States Environmental Protection Agency
EPA/ORIA	EPA's Office of Radiation and Indoor Air
ET	evapotranspiration
FEPs	features, events, and processes
GCL	geosynthetic clay liner
GM	geomembrane
GSI	Geosynthetic Institute
INL	Idaho National Laboratory
ISCOR	Interagency Steering Committee on Radiation Standards
LLC	Limited Liability Company
LLRW	low-level radioactive waste

LLW	low-level radioactive waste
NAS	National Academy of Sciences
NRC	United States Nuclear Regulatory Commission
NRC/FSME	NRC's Office of Federal and State Materials and Environmental Management Programs
NRC/RES	NRC's Office of Nuclear Regulatory Research
NUREG	Nuclear Regulatory Commission technical report designation
PA	performance assessment
QA/QC	Quality Assurance/Quality Control
RCRA	Resource Conservation and Recovery Act
RECAP	DOE/LM's Cover Renovation Research program
TCEQ	Texas Commission on Environmental Quality
UMTRA	Uranium Mill Tailings Remedial Action
UMTRCA	Uranium Mill Tailings Radiation Control Act
UNV	University of Nevada
USACOE	United States Army Corps of Engineers
USDA/ARS	United States Department of Agriculture's Agricultural Research Service
USGS	United States Geologic Survey
WIR	waste incidental to reprocessing

APPENDIX D: Bibliography

Albright, W. and Benson, C. (2009), *Store-and-Release Covers for Waste Containment: Principles and Practice*, ASCE Press, Reston, VA.

Albright, W., Benson, C., Gee, G., Abichou, T., McDonald, E., Tyler, S., and Rock, S. (2006). Field Performance of a compacted clay landfill final cover at a humid site. *Journal of Geotechnical and Geoenvironmental Engineering*, 132(11), 1393-1403.

Albright, W., Benson, C., Gee, G., Abichou, T., Tyler, S., and Rock, S. (2006). Field performance of three compacted clay landfill covers. *Vadose Zone Journal*, 5(4), 1157-1171.

Benson, C.H., W.H. Albright, D.O. Fratta, J.M. Tinjum, E. Kucukkirca, S.H. Lee, J. Scalia, P.D. Schlicht, and X. Wang, 2011 (in print). *Engineered Covers for Waste Containment: Changes in Engineering Properties and Implications for Long-Term Performance Assessment*, NUREG/CR-7028, Office of Research, U.S. Nuclear Regulatory Commission, Washington, DC.

Benson, C., Lee, S., Wang, X., Albright, W., and Waugh, W. (2008), Hydraulic Properties and Geomorphology of the Hydraulic Properties and Geomorphology of the Earthen Component of the Final Cover at the Monticello Uranium Mill Tailings Repository, Geo Engineering Report No. 08-04, University of Wisconsin, Madison, WI, USA.

Benson, C. and Meer, S. (2009), Relative abundance of monovalent and divalent cations and the impact of desiccation on geosynthetic clay liner, *Journal of Geotechnical and Geoenvironmental Engineering*, 133(5), 814-827.

Benson, C., Sawangsuriya, A., Trzebiatowski, B., and Albright, W. (2007), Post-construction changes in the hydraulic properties of water balance cover soils. *Journal of Geotechnical and Geoenvironmental Engineering*, 133(4), 349-359.

Benson, C.H., S.H. Lee, X. Wang, W.H. Albright, and W.J. Waugh, 2008. *Hydraulic properties and geomorphology of the earthen component of the final cover at the Monticello uranium mill tailings repository,* Geo Engineering Report No. 08-04, Geological Engineering, University of Wisconsin, Madison, Wisconsin.

Benson, C., Thorstad, P., Jo, H., and Rock, S. (2007). Hydraulic performance of geosynthetic clay liners in a landfill final cover, *Journal of Geotechnical and Geoenvironmental Engineering,* 133(7), 814-827.

Benson, C. and Wang, X. (2006), Temperature-Compensating Calibration Procedure for Water Content Reflectometers, *Proceedings TDR 2006: 3rd International Symposium and Workshop on Time Domain Reflectometry for Innovative Soils Applications,* Purdue University, West Lafayette, IN, USA, 50-1 - 5-16.

Benson, C., Waugh, W., Albright, W., and Smith, G. (2009), The RECAP Test Sections at the Grand Junction Disposal Site: Construction Documentation and Instrument Calibration, Geo Engineering Report No. 09-12, University of Wisconsin, Madison, WI.

Bradshaw, S. (2008), Effect of cation exchange during subgrade hydration and leachate permeation, MS Thesis, University of Wisconsin, Madison, WI.

Case, M. S., Case history of a 16-year old exposed HDPE geomembrane cover, IGS Conference Brazil, 2010 pp. 27.

DOE (U.S. Department of Energy), 2006, *Addendum 2 to the Performance Assessment for the Area 5 Radioactive Waste Management Site at the Nevada Test Site, Nye County, Nevada*, DOE/NV/11718--176-ADD2, Jun 2006

DOE (U.S. Department of Energy) 2009. *Technical Approach Document,* UMTRA-DOE/AL 050425.0002 Rev. 1, December.

DOE, Complex-Wide Review of DOE's Radioactive Waste Management, US Department of Energy, Office of Environmental Management, Washington, DC (2010).

EPA, September 2003, Evapotranspiration Landfill Cover Systems Fact Sheet, EPA 542-F-03-015, Office of Solid and Emergency Response, U.S. Environmental Protection Agency, Washington, DC [http://www.epa.gov/superfund/accomp/news/pdfs/evapo.pdf]

Garcia, C.A., Johnson, M.J., Andraski, B.J., Halford, K.J., and Mayers, C.J., 2008, Portable chamber measurements of evapotranspiration at the Amargosa Desert Research Site near Beatty, Nye County, Nevada, 2003–06: U.S. Geological Survey Scientific Investigations Report 2008-5135, 10 p. (Available at http://pubs.usgs.gov/sir/2008/5135/)

Garcia, C.A., Andraski, B.J., Stonestrom, D.A., Cooper, C.A., Johnson, M.J., Michel, R.L., and Wheatcraft, S.W., 2009, Transport of tritium contamination to the atmosphere in an arid environment: Vadose Zone Journal, v. 8, no. 2, p. 450-461, doi: 10.2136/vzj2008.0022.

Gish, T.J., and Kung, K.-J.S. 2007. Procedure for quantifying a solute flux to a shallow perched water table. Geoderma 138:57-64.

Guber, A. K., Pachepsky, Ya. A., van Genuchten, M. Th., Simunek, J., Jacques, D., Nemes, A., Nicholson, T. J., Cady, R. E. 2009. Multimodel simulation of water flow in a field soil using pedotransfer functions. Vadose Zone J 8:1–10.

Hanafy, S. and al Hagrey, S.. (2006). Ground-penetrating radar tomography for soil-moisture heterogeneity. *Geophysics.* 71(1): K9-K18

Kelln, C., Barbour, L., and Qualizza, C. (2009). Fracture-dominated subsurface flow and transport in a sloping reclamation cover. *Vadose Zone Journal*, 8, 96-107.

Koerner, R. (2005). *Designing with Geosynthetics*, 5th Ed., Prentice–Hall, Upper Saddle River, NJ.

Koerner, R., Hsuan, Y., and Koerner, G. (2005), Geomembrane Lifetime Prediction: Unexposed and Exposed Conditions, GRI White Paper No. 6, Geosynthetic Institute, Folsom, PA.

Letourneau, M.J., et al., The United States Department of Energy Radioactive Waste management Order 435.1 Update, The First Step – A Complex Wide Review, Proceedings from WM 2010, March 7-11, Phoenix, AZ (2010).

Letourneau, M.J., et al., Improving Consistency of Performance Assessments in the USDOE Complex, Proceedings from WM 2009, March 1-5, Phoenix, AZ (2009).

Lucius, J.E., Abraham, J.D., and Burton, B.L., 2008, Resistivity profiling for mapping gravel layers that may control contaminant migration at the Amargosa Desert Research Site, Nevada: U.S. Geological Survey Scientific Investigations Report 2008–5091, 30 p. (Available at http://pubs.usgs.gov/sir/2008/5091/)

Malusis, M. and Benson, C. (2006), Lysimeters versus Water-Content Sensors for Performance Monitoring of Alternative Earthen Final Covers, *Unsaturated Soils 2006*, ASCE Geotechnical Special Publication No. 147, 1, 741-752.

McGuire, P.E., Andraski, B.J., and Archibald, Ryan, 2009, Case study of a full-scale evapotranspiration cover: Journal of Geotechnical and Geoenvironmental Engineering, v. 135, no. 3, p. 316-332, doi:10.1061/(ACSE)1090-0241(2009)135:3.

McLendon, Terry, W. Michael Childress, Cade L. Coldren, Rick Frechette, and Frank Bergstrom. 2002. Evaluation of alternative cover designs for a water-balance cover over tailings at the Mineral Hill Mine, Montana, using the EDYS model. Tailings and Mine Waste 02. Proceedings of the Ninth International Conference on Tailings and Mine Waste. Balkema, Rotterdam. pp 505-518.

McLendon, Terry, Cade L. Coldren, and David L. Price. 2009. Comparison of results from the EDYS and EDYS-L ecological simulation models as applied to vegetation and hydrological dynamics on the Honey Creek Watershed, Texas. Technical Report ERDC TN-SWWRP-09-7. US Army Corps of Engineers. 19 p. [http://el.erdc.usace.army.mil/elpubs/pdf/swwrp-09-7.pdf]

Meer, S. and Benson, C. (2007). Hydraulic conductivity of geosynthetic clay liners exhumed from landfill final covers, *Journal of Geotechnical and Geoenvironmental Engineering*, 133(5), 550-563.

National Research Council (2007). *Assessment of the Performance of Engineered Waste Containment Barriers*, The National Academies Press, Washington, D.C.

NRC, "NRC Staff Guidance for Activities Related to U.S. Department of Energy Waste Determinations," NUREG-1854 (Draft Final Report for Interim Use), August 2007, ADAMS Accession No. ML072360184.

NRC, "Integrated Ground-Water Monitoring Strategy for NRC-Licensed Facilities and Sites: Logic, Strategic Approach and Discussion," NUREG/CR-6948, Volume 1, November 2007a.

Pachepsky Y.A., A.K. Guber, M.T. Van Genuchten, T.J. Nicholson , R.E. Cady, J. Simunek, M.G. Schaap, 2006. Model abstraction techniques for soil-water flow and transport, 65 NUREG/CR-6884, U.S. Nuclear Regulatory Commission, Washington, DC 20555- 0001.

Phifer, M. A., Crapse, K. P., Millings, M. R., and Serrato M. G. 2009. Closure Plan for the E-Area Low-Level Waste Facility, SRNL-RP-2009-00075. Savannah River National Laboratory, Aiken, South Carolina. March 16 2009.

Podgorney, R. and Bennett, J. (2006), Evaluating the long-term performance of geosynthetic clay liners exposed to freeze-thaw, *Journal of Geotechnical and Geoenvironmental Engineering*, 132(2), 265–268.

Rowe, R., Islam, M., Brachman, R., Arnepalli, D., and Ewais, A. (2010), Antioxidant Depletion from a High Density Polyethylene Geomembrane under Simulated Landfill Conditions, *Journal of Geotechnical and Geoenvironmental Engineering,* 136(7), 930–939.

Rowe, R., Rimal, S., and Sangam, H. (2009), Aging of HDPE geomembrane exposed to air, water, and leachate at different temperatures, *Geotextiles and Geomembranes*, 27, 137-151.

Rowe, R., Sangam., H.. and Lake, C. (2003). Evaluation of an HDPE geomembrane after 14 years as a leachate lagoon liner, *Canadian Geotechnical Journal*, 40, 536-550.

Scalia, J. (2009), Hydraulic and Chemical Properties of Geosynthetic Clay Liners Overlain by Geomembranes in Landfill Final Covers, MS Thesis, University of Wisconsin, Madison, WI.

Seitz, R.R. et al., Performance Assessment Community of Practice, Proceedings from WM 2010, March 7-11, Phoenix, AZ (2010).

Smesrud, J., Benson, C., Albright, W., Richards, J., Wright, S., Israel, T., and Goodrich, K. (2010), Using Pilot Test Data to Refine an Alternative Cover Design in Northern California, *International J. Phytoremediation*, in press.

SNL (Sandia National Laboratories), 2008, <u>Total System Performance Assessment Model/Analysis for the License Application</u>, MDL-WIS-PA-000005 Rev 00, AD 01, U.S. Department of Energy Office of Civilian Radioactive Waste Management

Stonestrom, D.A., Prudic, D.E., Walvoord, M.A., Abraham, J.D., Stewart-Deaker, A.E., Glancy, P.A., Constantz, J., Laczniak, R.J., and Andraski, B.J., 2007, Focused ground-water recharge in the Amargosa Desert Basin, *in* Stonestrom, D.A., Constantz, J., Ferré, T.P.A., and Leake, S.A., eds., Ground-water recharge in the arid and semi-arid southwestern United States: U.S. Geological Survey Professional Paper 1703, p. 107-136.

Walvoord, M.A., Andraski, B.J., Krabbenhoft, D.P., and Striegl, R.G., 2008, Transport of elemental mercury in the unsaturated zone from a waste disposal site in an arid region: Applied Geochemistry, v. 23, no. 3, p. 572–583, doi:10.1016/j.apgeochem.2007.12.014.

Waugh, W.J., C.H. Benson, and W.H. Albright, 2009. "Sustainable covers for uranium mill tailings, USA: Alternative design, performance, and renovation," Proceedings of 12th International Conference on Environmental Remediation and Radioactive Waste Management, Liverpool, UK.

Waugh, W.J., G.M. Smith, B. Danforth, G.W. Gee, V. Kothari, and T. Pauling, 2007. "Performance Evaluation of the Engineered Cover at the Lakeview, Oregon, Uranium Mill Tailings Site," Proceedings of Waste Management 2007 Symposium, Tucson, AZ.

Willgoose, G. R., 2005, *SIBERIA*, URL: http://www.telluricresearch.com/siberia-homepage.html

Yako, M. A, Case history of a 20-year old exposed HDPE surface impoundment liner, IGS Conference Brasil, 2010 pags 4.

Youngblood, J. H. "Long-Term Field Performance of HDPE Geomembranes in Landfill Applications," ASTM workshop, GRI-23 Conference on "Field Performance and Durability of Geosynthetics and Geosynthetic Related Systems," San Antonio, Texas, 2009 pp. 42.

APPENDIX E: Electronic Information Sources

American Society of Testing and Materials (ASTM)

http://www.astm.org/

ASTM D6766 - 09 Standard Test Method for Evaluation of Hydraulic Properties of Geosynthetic Clay Liners Permeated with Potentially Incompatible Liquids

http://www.astm.org/Standards/D6766.htm

ASTM Journal Article on In-Plane Hydraulic Properties of Geotextiles

http://www.astm.org/DIGITAL_LIBRARY/JOURNALS/GEOTECH/PAGES/GTJ10923J.htm

Desert Research Institute (DRI)

http://www.dr.edu/

DRI Center for Environmental Remediation and Monitoring

http://www.dri.edu/cerm

Radioactive Site Case Histories

Cornell Radiation Disposal Site (RDS)

http://www.eco.cornell.edu/DSC/rds.cfm

Long-Term Considerations for both the West Valley SDA and NDA

http://www.westvalleyeis.com/finaleis.htm

Niagara Falls Storage Site, Interim Waste Containment Structure (IWCS)

http://www.lrb.usace.army.mil/fusrap/index.htm

West Valley NRC-licensed Disposal Area (NDA)

http://www.wv.doe.gov/

West Valley State-licensed Disposal Area (SDA)

http://www.nyserda.org/programs/West_Valley

U.S. Army Corps of Engineer Headquarters

http://www.usace.army.mil/Pages/default.aspx

Ecological Dynamics Simulation (EDYS)

http://el.erdc.usace.army.mil/nrrdc/pdfs/edys-app.pdf

U.S. Department of Energy (DOE), Office of Environmental Management (EM)

http://www.em.doe.gov/Pages/EMHome.aspx

U.S. DOE Waste Sites and Locations

http://www.em.doe.gov/pages/siteslocations.aspx

3116 Waste Determinations Meeting and Long-term Engineered Cover Performance Issues Summary

http://www.em.doe.gov/pdfs/Long-TermEngineeredCapPerformance.pdf

U.S. DOE, Office of Legacy Management (LM)

http://www.lm.doe.gov/

U.S. DOE, Office of Legacy Management, Joint Environmental Management System (EMS)

http://www.lm.doe.gov/Office_of_Site_Operations/Environmental_Management_System.aspx

Regulatory Framework for DOE's Legacy Management Sites

http://www.lm.doe.gov/pro_doc/references/framework.htm

U.S. Environmental Protection Agency (EPA)

http://www.epa.gov/

U.S. EPA, Office of Solid Waste

http://www.epa.gov/osw/

U.S. EPA, Office of Solid Waste, Waste - Hazardous

http://www.epa.gov/wastes/hazard/

U.S. EPA, Office of Solid Waste, Landfills

http://www.epa.gov/osw/nonhaz/municipal/landfill.htm

U.S. Geological Survey, Amargosa Desert Research Site

http://water.usgs.gov/nrp/highlights/amargosa.html

U.S. Nuclear Regulatory Commission

http://www.nrc.gov/

U.S. NRC's Regulatory Guides

http://www.nrc.gov/reading-rm/doc-collections/reg-guides/

U.S. NRC's NUREG-Series Publications

http://www.nrc.gov/reading-rm/doc-collections/nuregs/

U.S. NRC's FSME Uranium Mill Tailings Website

http://www.nrc.gov/waste/mill-tailings.html

U.S. NRC's FSME Low-Level Waste Disposal

http://www.nrc.gov/waste/llw-disposal.html

U.S. NRC's FSME Low-Level Waste

http://www.nrc.gov/waste/low-level-waste.html

U.S. NRC's FSME Waste Incidental to Reprocessing

http://www.nrc.gov/waste/incidental-waste.html

World Information Service on Energy Uranium Project

http://www.wise-uranium.org/umtr.html

NRC FORM 335
(12-2010)
NRCMD 3.7

U.S. NUCLEAR REGULATORY COMMISSION

BIBLIOGRAPHIC DATA SHEET

(See instructions on the reverse)

1. REPORT NUMBER
(Assigned by NRC, Add Vol., Supp., Rev., and Addendum Numbers, if any.)

NUREG/CP-0195

2. TITLE AND SUBTITLE

Proceedings of the Workshop on Engineered Barrier Performance Related to Low-Level Radioactive Waste, Decommissioning, and Uranium Mill Tailings Facilities

held August 3 - 5, 2010 at the U.S. Nuclear Regulatory Commission Headquarters, Rockville, Maryland

3. DATE REPORT PUBLISHED

MONTH	YEAR
August	2011

4. FIN OR GRANT NUMBER

5. AUTHOR(S)

compiled by Thomas Nicholson and Hans Arlt

6. TYPE OF REPORT

Conference Proceedings

7. PERIOD COVERED *(Inclusive Dates)*

August 3 - 5, 2010

8. PERFORMING ORGANIZATION - NAME AND ADDRESS *(If NRC, provide Division, Office or Region, U.S. Nuclear Regulatory Commission, and mailing address; if contractor, provide name and mailing address.)*

Division of Risk Analysis, Office of Nuclear Regulatory Research, U.S. NRC, Mail Stop CSB-2A07, Rockville, MD 20852

9. SPONSORING ORGANIZATION - NAME AND ADDRESS *(If NRC, type "Same as above"; if contractor, provide NRC Division, Office or Region, U.S. Nuclear Regulatory Commission, and mailing address.)*

Division of Risk Analysis, Office of Nuclear Regulatory Research; and Division of Waste Management and Environmental Protection, Office of Federal and State Materials and Environmental Management Programs

10. SUPPLEMENTARY NOTES

11. ABSTRACT *(200 words or less)*

NRC's Offices of Nuclear Regulatory Research (RES) and Federal and State Materials and Environmental Management Programs (FSME) organized the Workshop on Engineered Barrier Performance Related to Low-Level Radioactive Waste, Decommissioning, and Uranium Mill Tailings Facilities. The workshop was held August 3-5, 2010 at NRC's Headquarters Auditorium. The workshop was coordinated with the States (i.e., Texas, South Carolina, Utah, Colorado, Washington, and New York), Tribal Nations (Navajo, Umatilla and Nez Perce), and Federal agencies (e.g., U.S. Department of Energy [DOE], U.S. Environmental Protection Agency [EPA], U.S. Department of Agriculture's Agricultural Research Service [USDA/ARS], U.S. Geological Survey [USGS], and DOE National Laboratories). The focus was engineered surface covers and bottom liners designed to isolate waste by impeding surface-water infiltration into the waste systems and mitigating the migration of contaminants from the waste disposal site. Topics included engineered barrier performance, modeling, monitoring, and regulatory experiences at low-level radioactive waste, decommissioning, and uranium mill tailings sites. The workshop objectives included: (1) facilitation of communication among Federal and State staff and contractors and selected experts on current engineered barrier issues and technical and regulatory experiences; (2) discussion of lessons learned and approaches for monitoring and modeling; (3) preparation of recommendations to address maintenance of engineered barrier performance over time; and (4) identification of topics for future research and the potential need to update technical guidance. Recommendations and insights given during session presentations, panel debates, and the discussions that followed were are included in this report.

12. KEY WORDS/DESCRIPTORS *(List words or phrases that will assist researchers in locating the report.)*

covers
decommissioning
engineered barriers
liners
low-level radioactive waste
model support
modeling
monitoring
performance assessment
radioecology

regulatory experiences
uranium mill tailings
water infiltration

13. AVAILABILITY STATEMENT

unlimited

14. SECURITY CLASSIFICATION

(This Page)

unclassified

(This Report)

unclassified

15. NUMBER OF PAGES

16. PRICE

Printed
on recycled
paper

Federal Recycling Program

UNITED STATES
NUCLEAR REGULATORY COMMISSION
WASHINGTON, DC 20555-0001

OFFICIAL BUSINESS

NUREG/CP-0195

Proceedings of the Workshop on Engineered Barrier Performance Related to Low-Level Radioactive Waste, Decommissioning, and Uranium Mill Tailings Facilities

August 2011

www.ingramcontent.com/pod-product-compliance
Lightning Source LLC
Chambersburg PA
CBHW080239180526
45167CB00006B/2339